Computational Studies, Nanotechnology, and Solution Thermodynamics of Polymer Systems

Computational Studies, Nanotechnology, and Solution Thermodynamics of Polymer Systems

Edited by

M. D. Dadmun

W. Alexander Van Hook

University of Tennessee
Knoxville, Tennessee

Donald W. Noid

Yuri B. Melnichenko

and

Bobby G. Sumpter

Oak Ridge National Laboratory
Oak Ridge, Tennessee

Kluwer Academic / Plenum Publishers
New York, Boston, Dordrecht, London, Moscow

Proceedings of Computational Polymer Science and Nanotechnology, and Solution Thermodynamics of Polymers, both held at the Southeastern Regional Meeting of the American Chemical Society, October 17–20, 1999, in Knoxville, Tennessee

ISBN 0-306-46549-3

233 Spring Street, New York, New York 10013

http://www.wkap.nl/

10 9 8 7 6 5 4 3 2 1

A C.I.P. record for this book is available from the Library of Congress

Printed in the United States of America

Preface

This text is the published version of many of the talks presented at two symposiums held as part of the Southeast Regional Meeting of the American Chemical Society (SERMACS) in Knoxville, TN in October, 1999. The Symposiums, entitled *Solution Thermodynamics of Polymers* and *Computational Polymer Science and Nanotechnology*, provided outlets to present and discuss problems of current interest to polymer scientists. It was, thus, decided to publish both proceedings in a single volume.

The first part of this collection contains printed versions of six of the ten talks presented at the Symposium on *Solution Thermodynamics of Polymers* organized by Yuri B. Melnichenko and W. Alexander Van Hook. The two sessions, further described below, stimulated interesting and provocative discussions. Although not every author chose to contribute to the proceedings volume, the papers that are included faithfully represent the scope and quality of the symposium.

The remaining two sections are based on the symposium on *Computational Polymer Science and Nanotechnology* organized by Mark D. Dadmun, Bobby G. Sumpter, and Don W. Noid. A diverse and distinguished group of polymer and materials scientists, biochemists, chemists and physicists met to discuss recent research in the broad field of computational polymer science and nanotechnology. The two-day oral session was also complemented by a number of poster presentations.

The first article of this section is on the important subject of polymer blends. M. D. Dadmun discusses results on using a variety of different co-polymers (compatiblizers) which enhance miscibility at the polymer-polymer interface. Following this article a series of papers are presented on the experimental production and molecular modeling of the structure and properties of polymer nano-particles and charged nano-particles (quantum drops). Related to this work is an article by Wayne Mattice on the simulation and modeling of thin films. The final paper included in this section is an intriguing article on identifying and designing calcium-binding sites in proteins.

The third section of the book presents an exciting selection of results from the current and emerging field of nanotechnology. The use of polymers for molecular circuits and electronic components is the subject of the work of P.J. MacDougall and J. A. Darsey. MacDougall *et al.* discuss a novel method for examining molecular wires by utilizing concepts from fluid dynamics and quantum chemistry. Another field of study represented in this section is the simulation of the dynamics of non-dense fluids, where, quite surprisingly, it was found that quantum mechanics might be essential for the study of nano-devices. Classical mechanical models appear to overestimate energy flow, and in particular, zero point energy effects may create dramatic instabilities. Finally, the article by R. E. Tuzun presents a variety of efficient ways to perform both classical and quantum calculations for large molecular-based systems.

The organizers are pleased to thank Professors Kelsey D. Cook and Charles Feigerle of the University of Tennessee, co-chairs SERMACS, for the invitations to organize the symposiums and for the financial support they provided to aid in their success. The organizers would also like to thank the Division of Polymer Chemistry of the American Chemical Society for financial support of the *Computational Polymer Science and Nanotechnology* symposium.

Mark D. Dadmun
W. Alexander Van Hook
Knoxville, TN

B.G. Sumpter
Don W. Noid
Yuri B. Melnichenko
Oak Ridge, TN

Symposium Schedule at SERMACS

Solution Thermodynamics of Polymers, I. - October 17, 1999

1. Solubility and conformation of macromolecules in aqueous solutions
   I. C. Sanchez, Univ. of Texas.

2. Thermodynamics of polyelectrolyte solutions
   M. Muthukumar, Univ. of Massachusetts.

3. Computation of the cohesive energy density of polymer liquids
   G. T. Dee and B. B. Sauer, DuPont, Wilmington.

4. Neutron scattering characterization of polymers and amphiphiles in supercritical carbon dioxide
   G. D. Wignall, Oak Ridge National Laboratory.

5. Static and dynamic critical phenomena in solutions of polymers in organic solvents and supercritical fluids
   Y. B. Melnichenko and coauthors, Oak Ridge National Laboratory.

Solution Thermodynamics of Polymers, II. - October 18, 1999

6. Nonequilibrium concentration fluctuations in liquids and polymer solutions
   J. V. Sengers and coauthors, Univ. of Maryland.

7. Polymer solutions at high pressures: Miscibility and kinetics of phase separation in near and supercritical fluids
   E. Kiran, Virginia Polytechnic Institute.

8. Phase diagrams and thermodynamics of demixing of polymer/solvent solutions in (T,P,X) space
   W. A. Van Hook, Univ. of Tennessee.

9. SANS study of polymers in supercritical fluid and liquid solvents
   M. A. McHugh and coworkers, Johns Hopkins University.

10. Metropolis Monte Carlo simulations of polyurethane, polyethylene, and betamethylstyrene-acrylonitrile copolymer
    K. R. Sharma, George Mason University.

Computational Polymer Science and Nanotechnology I – October 18, 1999

1. Pattern-directed self-assembly
   M. Muthukumar

2. Nanostructure formation in chain molecule systems
   S. Kumar

3. Monte Carlo simulation of the compatibilization of polymer blends with linear copolymers
   M. D. Dadmun

4. Atomistic simulations of nano-scale polymer particles
   B. G. Sumpter, K. Fukui, M. D. Barnes, D. W. Noid

5. Probing phase-separation behavior in polymer-blend microparticles: Effects of particle size and polymer mobility
   M. D. Barnes, K. C. Ng, K. Fukui, B. G. Sumpter, D. W. Noid

6. Simulation of polymers with a reactive hydrocarbon potential
   S. J. Stuart

7. Glass transition temperature of elastomeric nanocomposites
   K. R. Sharma

8. Stochastic computer simulations of exfoliated nanocomposites
   K. R. Sharma

Computational Polymer Science and Nanotechnology II – October 19, 1999

9. Simulation of thin films and fibers of amorphous polymers
   W. L. Mattice

10. Molecular simulation of the structure and rheology of lubricants in bulk and confined to nanoscale gaps
    P. T. Cummings, S. Cui, J. D. Moore, C. M. McCabe, H. D. Cochran

11. Classical and quantum molecular simulation in nanotechnology applications
    R. E. Tuzun

12. Conformational modeling and design of 'nanologic circuit' molecules
    J. A. Darsey, D. A. Buzatu

13. A synthesis of fluid dynamics and quantum chemistry in a momentum space investigation of molecular wires and diodes
P. J. MacDougall, M. C. Levit

14. Physical properties for excess electrons on polymer nanoparticles: Quantum drops
K. Runge, B. G. Sumpter, D. W. Noid, M. D. Barnes

15. Proton motion in SiO_2 materials
H. A. Kurtz, A. Ferriera, S. Karna

16. Designing of trigger-like metal binding sites
J. J. Yang, W. Yang, H-W. Lee, H. Hellinga

CONTENTS

THERMODYNAMICS OF POLYMER SYSTEMS

Phase Diagrams and Thermodynamics of Demixing of Polystyrene/Solvent Solutions in (T,P,X) Space 1
W. Alexander Van Hook

Thermodynamic and Dynamic Properties of Polymers in Liquid and Supercritical Solvents 15
Yuri B. Melnichenko, G. D. Wignall, W. Brown, E. Kiran, H. D. Cochran, S. Salaniwal, K. Heath, W. A. Van Hook and M. Stamm.

The Cohesive Energy Density of Polymer Liquids 29
G. T. Dee and Bryan B. Sauer

Thermal-Diffusion Driven Concentration Fluctuations in a Polymer Solution 37
J. V. Sengers, R. W. Gammon and J. M. Ortiz de Zarate

Small Angle Neutron Scattering from Polymers in Supercritical Carbon Dioxide 45
George D. Wignall

Polymer Solutions at High Pressures: Pressure-Induced Miscibility and Phase Separation in Near-Critical and Supercritical Fluids 55
Erdogan Kiran, Ke Liu and Zeynep Bayraktar

COMPUTATIONAL POLYMER SCIENCE

The Compatibilization of Polymer Blends with Linear Copolymers: Comparison between Simulation and Experiment 69
M.D. Dadmun

Nanoscale Optical Probes of Polymer Dynamics in Ultrasmall Volumes 79
M.D. Barnes, J.V. Ford, K. Fukui, B.G. Sumpter, D.W. Noid and J.U. Otaigbe

Molecular Simulation and Modeling of the Structure and Properties of Polymer Nanoparticles 93
B.G. Sumpter, K. Fukui, M.D. Barnes and D.W. Noid

Theory of the Production and Properties of Polymer Nanoparticles: Quantum Drops 107
K. Runge, K. Fukui, M. A. Akerman, M.D. Barnes, B.G. Sumpter and D.W. Noid

Simulations of Thin Films and Fibers of Amorphous Polymers 117
V. Vao-soongnern, P. Doruker and W.L. Mattice

Identifying and Designing of Calcium Binding Sites in Proteins by Computational Algorithm 127
W. Yang, H.-W. Lee, M. Pu, H. Hellinga and J.J. Yang

NANOTECHNOLOGY

A Synthesis of Fluid Dynamics and Quantum Chemistry in a Momentum-Space Investigation of Molecular Wires and Diodes 139
P.J. MacDougall and M.C. Levit

Classical and Quantum Molecular Simulations in Nanotechnology Applications 151
R. E. Tuzun

Computational Design and Analysis of Nanoscale Logic Circuit Molecules 159
K.K. Taylor, D.A. Buzatu and J.A. Darsey

Shock and Pressure Wave Propagation in Nano-fluidic Systems 171
D.W. Noid, R.E. Tuzun, K. Runge and B.G. Sumpter

Index 177

PHASE DIAGRAMS AND THERMODYNAMICS OF DEMIXING OF POLYSTYRENE/ SOLVENT SOLUTIONS IN (T,P,X) SPACE

W. Alexander Van Hook,

Chemistry Department
University of Tennessee
Knoxville, TN 37996-1600

ABSTRACT

Phase diagrams for polystyrene solutions in poor solvents and theta-solvents have been determined as functions of concentration, molecular weight and polydispersity, pressure, temperature, and H/D isotope substitution on both solvent and polymer, sometimes over broad ranges of these variables. The phase diagrams show upper and lower consolute branches and contain critical and hypercritical points. The isotope effects are large, sometimes amounting to tens of degrees on critical demixing temperatures. Most often solvent deuterium substitution decreases the region of miscibility and substitution on the polymer increases it. The demixing process has been investigated using dynamic light scattering (DLS) and small angle neutron scattering (SANS), examining pressure and temperature quenches from homogeneous conditions to near-critical demixing. The SANS and DLS results (which refer to widely different length scales) are discussed in the context of scaling descriptions of precipitation from polymer solutions.

INTRODUCTION

Liquid-liquid (LL) demixing of weakly interacting polymer/solvent solutions such as polystyrene(PS)/ acetone(AC), PS/ cyclohexane(CH), PS/methylcyclohexane(MCH), etc. is characterized in the temperature/ segment-fraction (T,ψ) plane by the presence of upper and lower demixing branches.[1-4] Some solvents dissolve some polymers at all accessible temperatures (*i.e.* between the melting point of the solvent and its liquid/vapor critical point), no matter the length of the chain. These are the so called "good solvents", and the solutions, while viscous and perhaps hard to handle, are homogeneous across the entire concentration range, $(0 \leq \psi \leq 1)$, *e.g.* polystyrene(PS) in benzene. Other solvents (*e.g.* CH, MCH) dissolve very long chains (in the limit, infinitely long chains) for $(0 \leq \psi \leq 1)$, but only within a limited range of temperature $|(T_{\Theta,U} \leq T \leq T_{\Theta,L})|$. Here $T_{\Theta,U}$ and $T_{\Theta,L}$ are the upper and lower Flory Θ-

temperatures, respectively. Finally there exists a class of poor solvents which are unable to dissolve long polymer chains (and in some cases even short ones) at any appreciable concentration, A good example is PS/acetone. Acetone does dissolve short chain PS, but the limit (192 monomer units at the critical concentration) is small enough to destroy the utility of this solvent in all but special cases.[5]

DISCUSSION

Phase equilibria in monodisperse and polydisperse polymer solutions: The discussion of demixing from polymer solutions can be simplified by considering Figure 1. This figure shows the most common type of phase diagram for PS/solvent mixtures in (ψ,T,X) space, (ψ = segment fraction PS, T=temperature, X some third variable of interest). To begin, consider a solution held at constant pressure (nominally 1 atm.), and let X scale with molecular weight (M_W). Flory-Huggins theory suggests $X=M_W^{-1/2}$, and as expected, the extent of the one phase homogeneous region increases with X (*i.e.* long chain polymers are less soluble than short-chain ones). For solutions in θ-solvents (Figure 1a) an extrapolation of the heavy line drawn through the maxima or minima of the consolute curves (which in first approximation coincide with the upper and lower critical points) yields X=0 values (*i.e.* intercepts at infinite M_W) defining the upper and lower θ-temperatures. By general acceptance the term "upper critical temperature" or "upper consolute temperature", UCS, refers to that part of the demixing diagram with $(\partial^2T/\partial\psi^2)_X<0$, while "lower critical" or "lower consolute temperature", LCS, refers to $(\partial^2T/\partial\psi^2)_X>0$. For diagrams such as Figure 1 it follows that $T_{\Theta,U} < T_{\Theta,L}$. In Figure (1a) the upper and lower $(T,X)_{\psi CR}$ curves have been connected using an empirical smoothing function (the dotted line) which extends into a hypothetical region, X<0.[2]

Figure 1b represents demixing from a poor solvent. Here the UCS and LCS branches join at a double critical (or "hypercritical") point, this time located at X>0 (*i.e.* at real MW). Continuing, one might argue that the principal feature which distinguishes demixing from θ-solvents and poor-solvents (Figure 1a from 1b) is nothing more than a shift of the diagram along the X coordinate. In poor solvents the $(T,X)_{\psi cr}$ projection displays its extremum (or hypercritical point, $(\partial X/\partial T)_{\psi CR}=0$ and $(\partial^2X/\partial T^2)_{\psi CR}>0$, designated a lower hypercritical temperature $T_{HYP}{}^L$) at real X, (*i.e.* $X_{HYP}>0$). For $X<X_{HYP}$ the system collapses into the hour glass configuration (see the darkest shading in Figure 1b). In contrast, solutions in θ-solvents show extrema at X<0, *i.e.* at P<0 in the $(T, \psi, X=P)_{MW>0}$ projection (perhaps experimentally inaccessible and perhaps hypothetical), or at negative $M_W^{-1/2}$ in the $(T,\psi,X=M_W^{-1/2})_{P>0}$ projection (definitely inaccessible and certainly hypothetical).

The discussion above has described precipitation from solutions of monodisperse polymers where the M_W is well defined and the LL demixing diagram is constrained to lie on one or another of the shaded planes in Figure 1. Often, however, it is necessary to account for the M_W fractionation which occurs on precipitation because of polydispersity in the polymer sample. Figure 1c diagrams that situation. The parent phase of (average) M_W and concentration $(\langle M_W\rangle, \psi)_A$ at point A, is in equilibrium with daughter phase of somewhat higher concentration and larger $\langle M_W\rangle$, at, say, point C. The tie line which connects the polymer-poor parent, $(T_A, \psi_A, \langle M_W\rangle_A)$, and polymer-rich daughter phases, $(T_C, \psi_C, \langle M_W\rangle_C)$, does not lie in the $(T, \psi)_{MW-A}$ plane, but rather angles across the (ψ, X) projection. Similarly, the equilibrium between polymer-rich parent $(T_B, \psi_B, \langle M_w\rangle_B)$ and polymer-poor daughter phases $(T_D, \psi_D, \langle M_W\rangle_D)$ to the other side of the diagram also skews across (ψ, X), but at a different angle. Given a sufficiently detailed expression which defines the equation of state for the solution, G(T, ψ, X), G the Gibbs free energy, the equilibrium surface defining parent daughter equilibrium can be constructed. The parent phases, A, B,....*etc*, define the cloud-point surface, CP, which lies at a constant value of X, while the daughter phases, C, D,..... *etc*. lie on the shadow curve, SHDW, which is skewed with respect to X (the skewing angle being

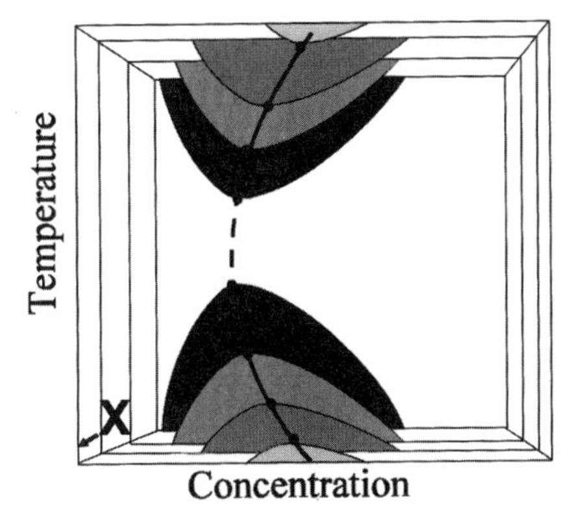

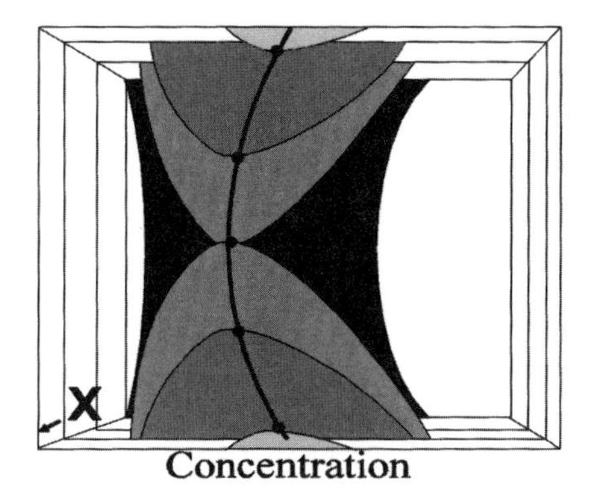

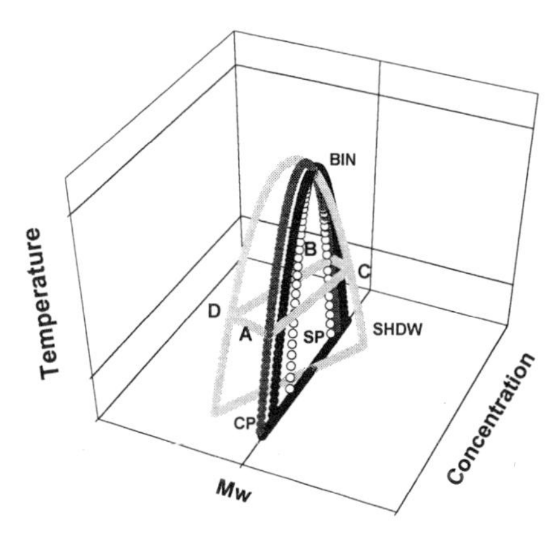

Figure 1. Demixing diagrams for PS in θ-solvents and poor solvents (schematic). The variable X might be pressure, $M_w^{-1/2}$, D/H ratio in solvent or solute, etc. See text for a further discussion. (a, top left) PS in a θ-solvent (monodisperse approximation). For $X=M_w^{-1/2}$ the X=0 intercepts of the upper and lower heavy lines drawn through the minima or maxima in the demixing curves define θ_L and θ_U, respectively. (b, top right) PS in a poor solvent (monodisperse approximation). The heavy dot at the center locates the hypercritical (homogeneous double critical) point. (c, bottom right) The effect of polydispersity. BIN=binodal curve, CP=cloud point curve, SP=spinodal, SHDW=shadow curve. See text. Modified from ref. 6 and used with permission.

determined by the extent of polydispersity. The critical point, $(T_{CR}, \psi_{CR})_X$, is assigned to the intersection of CP and SHDW, and at this point the spinodal curve, SP, is tangent to CP. SP defines the limit of metastability for demixing and is obtained from the loci of points of inflection on the $(G, T, \psi)_{X_{parent}}$ surface. The demixing curve in the monodisperse approximation, BIN is also shown in the Figure. All four curves, CP, SHDW, SP and BIN, are common at $(T_{CR}, \psi_{CR})_{<X>}$. For monodisperse samples, BIN, CP and SHDW coincide. Luszczyk, Rebelo and Van Hook[6] have developed a mean-field formalism and computational algorithims which interpret CP and SP data on LL demixing, explicitly considering effects of P, T, ψ, M_W, polydispersity, and H/D substitution on the parameters defining the free energy surface.

Polymer phase equilibria at positive and negative pressures: So much for projections in $(T,\psi,X=M_W^{-1/2})_P$ space. We next consider demixing in one or another $(T,\psi,X=P)_{Mw}$ projection, *i.e.* by first fixing M_W at a convenient value, then measuring demixing curves in the $(T,\psi)_{Mw}$ plane at various pressures. Most commonly as P=X increases, moving out from the plane of the paper (Figure 1), solvent quality improves. In such a state of affairs it is possible to select initial values of solvent quality, P, T, and M_W so that the solution lies in the one-phase homogeneous region but not too far from the LL equilibrium line. Demixing is induced by quenching either T or P. Depending on the precise shape of the diagram and the specific starting location this may be accomplished by either raising or lowering T, raising or lowering P, or by a combination of changes. Of course if the solvent be poor enough, one can force precipitation by increasing M_W, or modifying solvent quality (for example by isotope substitution), but these are variables we agreed to hold constant in this first part of the discussion. By especially careful choice of solvent quality, T, and M_W, one can locate the one-phase mixture at P~0 such that further lowering the pressure (to negative values, *i.e.* placing the solution under tension) induces precipitation. This assumes the equation of state describing the solution is well behaved and continuous across P=0, and smoothly extends into the tensile region. An example is discussed below.

To simplify the discussion we make an additional abstraction and consider projections from $(P,T,\psi,X=M_W^{-1/2})$ space onto a three dimensional critical surface $(P,T,X=M_W^{-1/2})_{\psi CR}$ by holding the concentration at its critical value, see Figure 2. In Figure 2 we show $(T,P)_{Mw,\psi cr}$ sections at two M_W's, and $(T,X=M_W^{-1/2})_{P,\psi cr}$ sections at four pressures, including two for P>0, and one for P<0. The $(T,P)_{MW=\infty,\psi crit}$ projection at the left refers to $X=M_W^{-1/2}=0$. It maps the pressure dependence of Θ_{UCS} and Θ_{LCS} in the $(T,P)_{\psi cr,X=0}$ plane. In this figure we chose $\Theta_{UCS} = \Theta_{LCS} = T_{HYP}^L$ at P=0 which, while certainly possible, requires careful tuning of solvent quality. In this example the solvent is to be labeled as a Θ-solvent for P>0, and a poor solvent for P<0. The $(T,P)_{MW=\infty,\psi crit}$ projection to the right is similar, but this time maps LL equilibria at some finite M_W (X>0). The insert sketches two possible shapes for the master curve which describes demixing in the $(T,P)_{MW,\psi crit}$ plane[4,7] and is applicable to type III, IV, and V diagrams in the Scott-von Konynenburg classification.[8] Although curvature in (T,P) plots is not thermodynamically required our interest is in systems such as PS/methylcyclohexane and PS/propionitrile) where it is found.

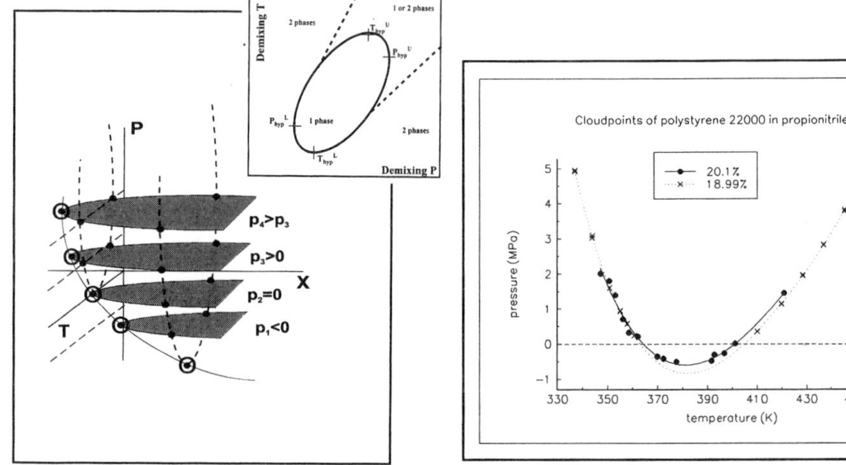

Figure 2 (left). A critical demixing diagram in $(T, P, X=M_w^{-1/2})_{\psi CR}$ space. Isobars at P>0, P=0, and P<0 are shown. In this schematic $\Theta_L=\Theta_U$ at P=0 which demands careful tuning of solvent quality. Isopleths for X=0 and X>0 are shown. The insert is a (schematic) isopleth at X>0. Several possible behaviors in the region of high pressure and high temperature are shown, see text for further discussion. Modified from ref. 1 and used with permission.

Figure 3 (right). Continuity of a demixing isopleth at negative pressure. The demixing isopleth of PS(22,000) in propionitrile. (See text). Modified from ref. 1 and used with permission.

The master curve shows at least two hypercritical points, P_{HYP}^L and T_{HYP}^L, characterized by $(\partial P/\partial T)_{CRIT}=0$ and $(\partial^2P/\partial T^2)_{CRIT}>0$, and $(\partial T/\partial P)_{CRIT}=0$ and $(\partial^2T/\partial P^2)_{CRIT}>0$, respectively. Numerous examples of systems with either P_{HYP}^L or T_{HYP}^L, but not both, have been reported, and we have recently reported[8] the first example of a binary polymer/solvent mixture which shows both P_{HYP}^L or T_{HYP}^L (several examples of binary mixtures of small molecules exhibiting both hypercritical points have been discussed by Schneider[9]). The shape of the master curve in the region toward high T and high P is not established. We have been unable to find reports of experiments in these regions, but simulations suggest that the high temperature part of the lower branch (i.e. the section to the high temperature side of P_{HYP}^L) turns back to lower pressure, after reaching a maximum[10, 11] and that is one behavior sketched in the insert to Figure 2. The other possibility which shows a closed one-phase loop as an island in a two-phase sea is more speculative, but is included as an interesting possibility. That possibility contains upper and lower hypercritical temperatures, T_{HYP}^U and T_{HYP}^L, and upper and lower hypercritical pressures, P_{HYP}^U and P_{HYP}^L. Open and closed reentrant phase diagrams like those illustrated in Figures 1 and 2 have been discussed by Narayanan and Kumar[12] and are discussed or implied in the developments of Schneider,[13] Prigogine and Defay,[14] and Rice.[15] From such analyses we have concluded[16] that any simple Flory-Huggins model leading to a closed loop in the $(T,P)_{MW,\psi crit}$ projection must include T and P dependent excess free energy (χ) parameters.[6] A simple equation which predicts significant curvature in the (T,P) plane (in the limit a closed loop), and which satisfies all relevant thermodynamic constraints, is found when the χ parameters describing excess volume and excess enthalpy are each dependent on T and P, but in compensatory fashion.

SOME EXAMPLES

a. Continuity at negative pressure (Θ-solvent/poor-solvent transitions). Imre and Van Hook used the Berthelot technique to generate negative pressures in order to induce phase transitions in some different polymer/solvent systems.[17] In PS/propionitrile, PS/PPN, PPN a poor solvent, they demonstrated continuity for the demixing curve across P = 0 and well into the region P<0. It is the choice of solvent quality which dictates whether the hypercritical point lies at P>0, P~0, or P<0. In designing experiments at negative pressure (including the choice of solvent and polymer M_w) one is strictly limited to tensions which are smaller than the breaking strength (cavitation limit) of the liquid itself, or the adhesive forces joining liquid to wall. Figure 3 shows CP data for a 0.20 wt. fraction PS (M_w=22,000) over the range (2>P/MPa>-1), comparing those results with values at higher pressure obtained by another technique.[18] The two data sets agree nicely along both UCS and LCS branches and confirm that the equation of state for this solution passes smoothly and continuously across zero pressure into the region of negative pressure. The authors concluded that it is physically reasonable to compare properties of solutions at positive and negative pressure using continuous and smoothly varying functions. For example it may be convenient to represent an isopleth (including the critical isopleth) in terms of an algebraic expansion about the hypercritical origin, even when that origin is found at negative pressure. Such expansions have been found to be useful representations of demixing even when the hypercritical origin lies so deep as to be experimentally inaccessible, or is below the cavitation limit.[2]

In a related study on PS/methyl acetate (PS\MA) we[19] examined the θ-solvent/poor-solvent transition at negative pressure (refer to the discussion around Figure 2). MA is a θ-solvent at ordinary pressure and the transition corresponds to a merging of the UCS and LCS branches at negative pressure. For PS of M_w =2x10^6 the hypercritical point lies below -5 MPa and was experimentally inaccessible (as it was for M_w =2x10^7). However CP measurements were carried out at pressures well below P=0 thus establishing continuity of state and showing the likely merging of the UCS and LCS branches.

The importance of experiments at negative pressure is that they establish continuity of state across the P=0 boundary into the region where solutions are under tension. In this line of thinking the UCS and LCS demixing branches share common cause. That interpretation forces a broadening of outlook which has been useful. For example, an immediate and practical extension was the development of a scaling description of polymer demixing in the $(T,X=M_w^{-1/2})_{\psi cr,P}$ plane.[2] That description employs an expansion about the hypercritical origin, X_{HYP}, even for X_{HYP}<0. The approach is in exact analogy to expansions about P_{HYP} (whether positive or negative).

b. Marked curvature for critical demixing in the $(T,P)_{MW,\psi crit}$ projection. Two component and one component solvents. Although weakly interacting polymer/solvent systems showing T_{HYP}^L **or** P_{HYP}^L (but not both) have been long known, it was not until recently that the pace of experimental work increased to the point where detailed comparisons of theory and experiment became possible. We wanted to find weakly interacting systems with sufficient curvature to display both T_{DCP}^L **and** P_{DCP}^L, partly because such systems would afford a good test of commonly used thermodynamic and/or theoretical descriptions of weakly interacting polymer solutions. Interest in scaling descriptions of thermodynamic properties and of intensities of light and neutron scattering during the approach to the critical isopleth further encouraged the search.

In looking for a system with two double critical points we examined a series of two and three component systems.[7] For two component studies we chose solutions showing significant curvature in the $(T,P)_{crit}$ projection, usually with known T_{HYP}^L or P_{HYP}^L at convenient M_W. Unfortunately, in each case the curvature was insufficient to display both T_{HYP}^L and P_{HYP}^L within experimentally accessible ranges, $(\sim 270 < T/K < \sim 500)$ and $(-1 < P/MPa < 200)$. For example, PS solutions of various M_w dissolved in the θ-solvents CH or MCH show well defined T_{HCP}^L at reasonable T and P, but the pressure dependence is such that if P_{HYP}^L occurs at all it lies at too deep a negative pressure to be observed. Interestingly, solutions of PS in the commercially available mixture *(cis:trans::1:1)*-dimethylcyclohexane($DMCH_{Cis/Trans//1/1}$) show significantly more curvature but still not enough to display both P_{DCP}^L and T_{DCP}^L (but we will return to PS/DMCH solutions below). Neither did we have success in studies of PS dissolved in other poor solvents. Both PS/acetone and PS/propionitrile show well developed P_{HYP}^L at P~0.1 MPa and convenient values of T and M_w, but increasing the pressure to 200 MPa fails to develop T_{HYP}^L.

c. A PS/(two-component solvent) mixture with two hypercritical points. In two component solvents one hopes that mixing two solvents (typically a θ-solvent and a poor-solvent), each with conveniently located T_{HYP}^L or P_{HYP}^L, will result in a solution with both extrema. Preliminary experiments on PS/(cyclohexane (CH)+propionitrile(PPN)) and PS/(methylcyclohexane (MCH)+ acetone(AC)) systems were unsuccessful, but trials on PS/n-heptane/MCH system where polymer/solvent interaction is nonspecific, showed both T_{HYP}^L and P_{HYP}^L (Figure 4). In the discussion of Figure 4 we assume ψ_{crit}, for PS($MW=2.7\times10^6$) in HE/MCH mixtures is independent of HE/MCH ratio, and equal to its value in MCH. This point of view is supported by Flory-Huggins theory which suggests for noninteracting solutions "the main contribution of the solvent is primarily that of lowering the critical solution temperature by dilution. The exact nature of the solvent is of only secondary importance" (R. L. Scott[20]).

The rationale for studying CPC's in the mixed solvent HE/MCH system followed from first order FH analysis which argues that modest decreases in solvent quality are expected to raise P_{HYP}^L toward higher temperature and T_{HYP}^L to higher pressure. The data in Figure 4 show this to be correct. HE is a much poorer solvent than MCH and the shift in solvent quality from MCH to HE/MCH (0.2/0.8) shift P_{HYP}^L and T_{HYP}^L significantly. Both double critical points are now observed in the range $(0MPa < P < 200MPa)$, (inserts to Figure 4).

d. A PS/(one-component solvent) mixture with two hypercritical points. The practical possibility of demixing curves with both P_{HYP}^L and T_{HYP}^L established, we reconsidered the PS/1,4-DMCH system. According to Cowie and McEwen[21] a 1:1 mixture of *cis/trans* isomers of 1,4-DMCH is a poor solvent for PS, but our preliminary measurements on samples of intermediate M_w failed to confirm that observation, and, continuing, we compared PS solubility in mixed and unmixed *trans*-and cis-1,4-DMCH, finding the *trans* isomer to be the worse solvent. The best chance, then, of observing multiple hypercritical points should be in

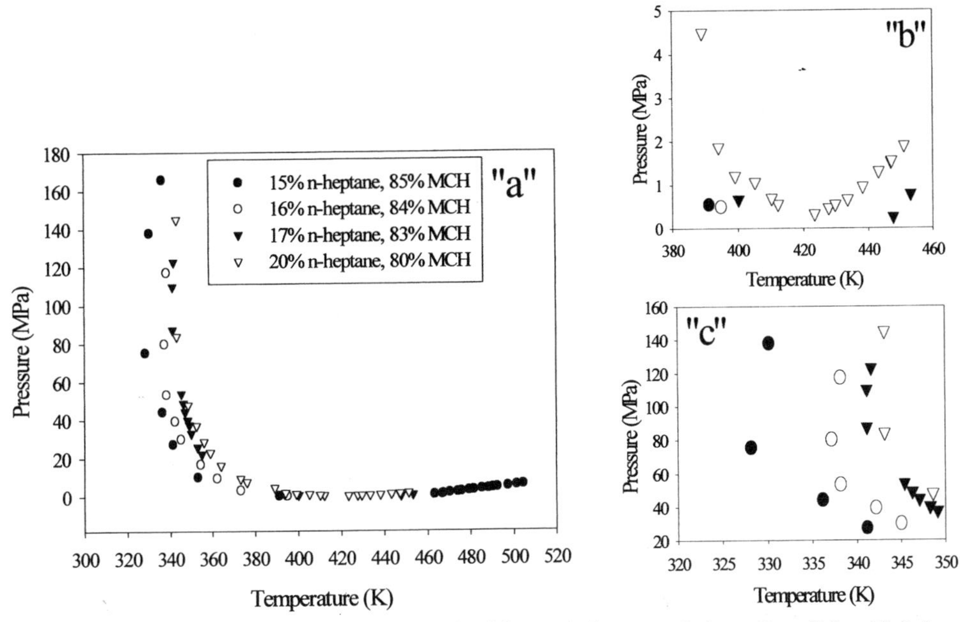

Figure 4. Critical Demixing isopleths for PS/methylcyclohexane/n-heptane solutions. Parts "b" and "c" show the diagrams in the vicinity of the hypercritical (homogeneous double critical) points. Modified from ref. 4 and used with permission.

the poorer solvent, *trans*-1-4-DMCH, but with M_W carefully chosen to properly size the one phase homogeneous region. For PS9x10^5 $T_{HYP}{}^L$ lies slightly above 200 MPa but for this solution $P_{HYP}{}^L$<0. Therefore M_w was decreased slightly to yield $T_{HYP}{}^L$ for (PS5.75x10^5(7wt%)/trans-1,4-DMCH) at 175 MPa and 349.15 K. For this solution $P_{HYP}{}^L$ lies at P=1.65MPa, 438.7K. This set of measurements establishes that the proposed master curve exists in at least one weakly interacting binary polymer solution (see Figure 5).

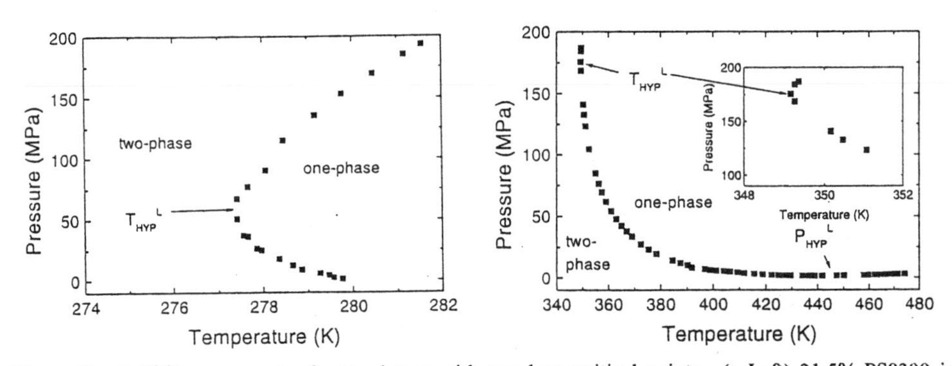

Figure 5. A PS/1-component solvent mixture with two hypercritical points. (a Left) 21.5% PS8300 in cis/trans//1/1-1,4-dimethylcyclohexane(DMCH) exhibiting only $T_{HYP}{}^L$ in this experimental range. (b Right) 7% PS575,000 in trans-1,4-DMCH showing both $T_{HYP}{}^L$ and $P_{HYP}{}^L$. Modified from ref. 7 and used with permission.

e. A reduced description of curvature in the (T,P) demixing plane. To our knowledge the examples above constitute the only weakly interacting polymer/solvent systems now known with two hypercritical points (homogeneous double critical points). To facilitate comparisons with other experiments or theory it is useful to employ fitting equations containing the minimum set of parameters. In the present case polynomial expansions are inconvenient because $(P,T)_{CP}$ loci in some regions are double valued. Higher order terms are required and the fits are no longer economical so far as number of parameters is concerned. We therefore elected rotation to a new coordinate system, (π,τ), observing that in the new system the demixing data set is symmetrically disposed about a single extremum. The transformation equations are

$$\tau = [T^2 + P^2]^{1/2}\cos\{\arctan(P/T) + \alpha\} \quad \text{and} \quad \pi = [T^2 + P^2]^{1/2}\sin\{\arctan(P/T) + \alpha\} \quad (1)$$

where α is the angle of rotation and the new minimum is selected by a minimization routine. An logarithmic expansion centered at that minimum results in a scaling fit which is characterized by an economy of parametrization.

$$|(\tau - \tau_{min})/\tau_{min}| = A\,|(\pi - \pi_{min})/\pi_{min}|^{\nu} \quad (2)$$

In Equation 2, A is a width factor, ν an exponent, τ and π are the transformed (T,P) coordinates, and τ_{min} and π_{min} are coordinates of the new origin. Least squares parameters for the plots in Figures 4 and 5 are reported in Table 1. Figure 6a is a logarithmic representation of the PS/MCH/HE data. It is linear over ~2 ½ orders of magnitude. Figure 6b compares that least squares fit, now transformed from the (π,τ) coordinate system back to (P,T), with experiment. The quality of fit to the PS/*trans*-1-4-DMCH data (not shown) is comparable. (For either set of solutions the use of a cubic polynomial in place of Equation 2 fails to represent the data set within experimental error although the number of parameters is the same.) The symmetry exhibited by the $(T,P)_{crit}$ loci after transformation to (π,τ) space is a point of special interest. It will be important to determine whether the exponent which describes curvature along $(\pi,\tau)_{CR}$, $\nu = 0.61 \pm 0.04$, will carry over to other polymer/solvent systems, and to find whether the parameter α correlates with other thermodynamic properties of solution. It is interesting that the purely empirical scaling exponent describing the $(\pi,\tau)_{CR}$ isopleth is numerically equal to the theoretically established scaling exponent which describes divergences in DLS and SANS correlation length and intensity on critical demixing.

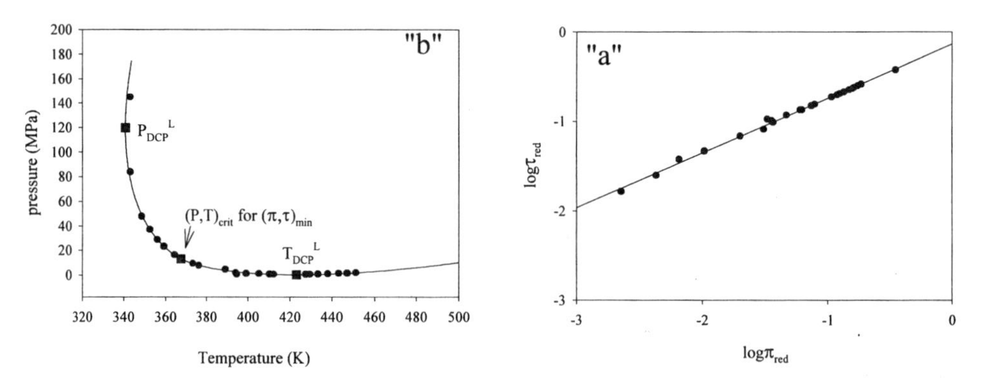

Figure 6. Reduced representation of critical demixing isopleths of high curvature. The PS/MCH/HE system, (a) logarithmic representation in the (π,τ) plane. (b) Same as (a) but after transformation back to (P,T) coordinates, see Equations (1) and (2), Figure 4, and text. Modified from ref. 4 and used with permission.

f. The effect of H/D substitution. Curvature in (P,T,y_D) space. We now turn to H/D substitution on solvent or polymer. It is now clear from data on PS/acetone(AC),[5,6]

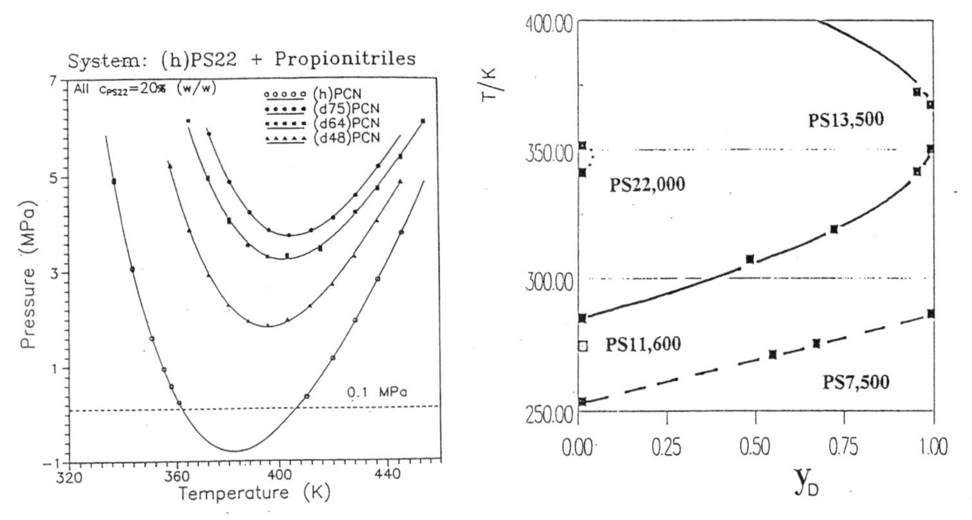

Figure 7. (a) Solvent isotope effects on demixing of PS/propionitrile mixtures, CH_3CH_2CN/CH_3CD_2CN. (b) Demixing of PS/acetone mixtures at various M_w's and isotope ratios, $(CH_3)_2CO/(CD_3)_2CO$. Notice the increased curvature as the hypercritical M_w is approached. Modified from refs. 6 and 18 and used with permission.

PS/propionitrile(PPN),[18] and PS/MCH [16,22] solutions that the demixing isopleths are very sensitive to solvent quality and such sensitivity extends to effects of H/D substitution. Several examples follow.

Figure 7a shows demixing isopleths for PS/PPN solutions as a function of D substitution on PPN in the methylene position. As $y_{D,CH2}$ increases 0.00 to 0.75 ($y_{D,CH2}$= solvent fraction methylene deuteration, and z_D = polymer fraction deuterated), $P_{HYP}{}^{L}$ shifts markedly and nonlinearly from its origin at (383K, -1.0MPa, y_D =0.00) to (402K, 3.8MPa, y_D=0.75). (Although not shown on this figure, it is this $y_{D,CH2}$=0 solution which was studied under tension to establish continuity of state for demixing at P<0 (Figure 3)). In Figure 7b we turn attention from curvature of the demixing isopleths at selected values of y_D, and show curvature in the $(T,y_D)_{\psi CR}$ plane at several M_w for PS/AC mixtures. The figures are analogous to those described in the (T,P) plane, above, and mean-field or scaling descriptions of isotope effects on the demixing loci follow directly.[6]

DYNAMIC LIGHT SCATTERING (DLS). AND SMALL ANGLE NEUTRON SCATTERING (SANS) NEAR CRITICAL DEMIXING ISOPLETHS.

The development above gives examples of thermodynamic information which can be extracted from studies of the T, P, M_w, y_D and z_D dependences of critical demixing. More recently we turned attention from the purely thermodynamic description of LL transitions to studies of the mechanism of LL precipitation employing dynamic light scattering (DLS) and small angle neutron scattering (SANS). It is interesting to determine scattering intensities and correlation lengths near LL transitions, especially in regions of high curvature, because such measurements permit exploration of the proper thermodynamic path to employ in multidimensional scaling descriptions of the approach to criticality.[23] The example shown in Figure 8 reports DLS and SANS data for PS30k in $MCHh_{14}$ (11.5 wt%, 12.1 segment%) and $MCHd_{14}$ (11.4 wt%, 13.3 segment%). Although referring to slightly different concentrations the (T,ψ) curves are flat near the maximum and both solutions are near critical. The data nets are shown in Figure 8a. We chose deuterated solvent to permit direct comparison with SANS data. Deuteration is widely employed in SANS to set the contrast. Figures 8b and 8c show

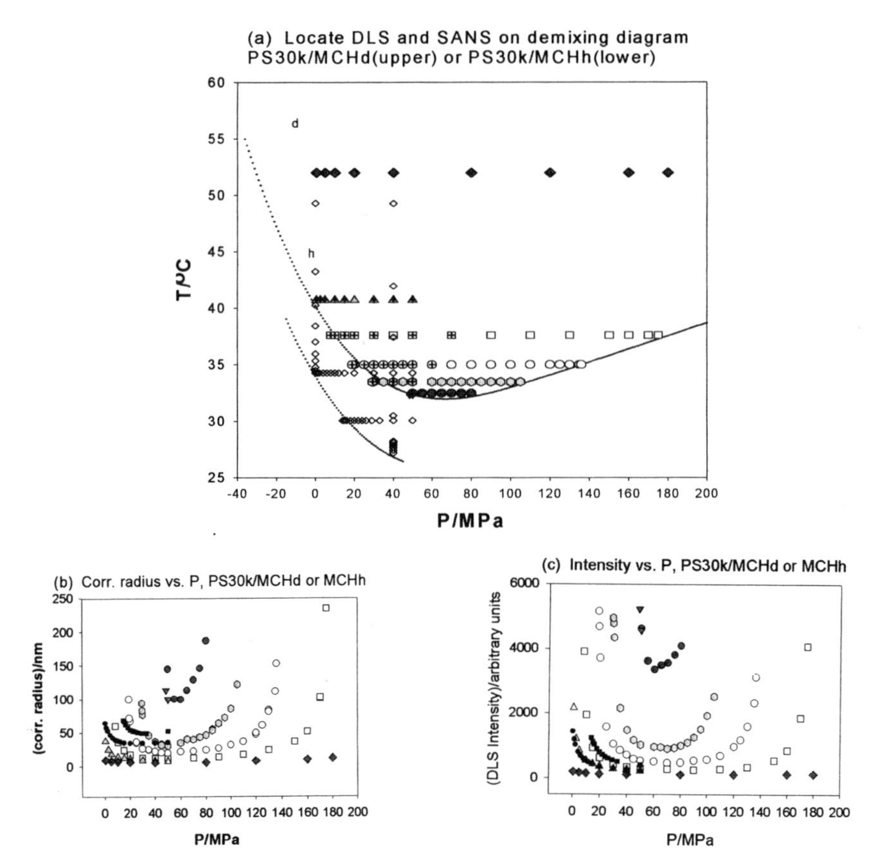

Figure 8. (a) DLS and SANS investigations in the $(T,P)_{\psi CR}$ plane for PS30k/MCHh and PS30k/MCHd solutions. Open and shaded symbols = DLS, Closed and centered symbols = SANS. (b,c) DLS and SANS correlation radii and scattering intensity at several isotherms for the data shown in part (a). Notice the divergences observed along several isotherms as pressure is either raised or lowered toward the critical line. Modified from ref. 16 and used with permission.

DLS and SANS correlation lengths, ξ, and scattering intensities, I, for PS30k/MCHd and PS30k/MCHh. Divergences in ξ and I are found as the critical line is approached by increasing or decreasing P, or by decreasing T. Similar data is available at other M_w's in MCH and for a variety of other PS/solvent systems.[16,22]

A scaling representation of DLS and SANS correlation radii. In a review of reentrant LL phase transitions, Narayanan and Kumar[12] comment, "...the most compelling issue concerning experimental investigations of reentrant phase transitions (RPT's) is how to recover universal exponents for the RPT. In other words which field variable should be used in place of ....?", and elsewhere, "In other words, it is difficult to obtain the correct thermodynamic path in the multidimensional field space." In common with others they demonstrate that a doubling of critical exponents follows from the geometrical model of RPT's for data sufficiently close to the critical line (provided that line be quadratic in the vicinity of the hypercritical turning point). Near P_{HYP}^{L} or P_{HYP}^{U} it is a properly defined reduced temperature which should demonstrate the doubling, near T_{HYP}^{L} or T_{HYP}^{U} it is a properly reduced pressure. They remind us that the properly defined scaling variable can nowhere be tangent to the critical curve.

The present data extend over a wide range and in no sense carefully sample the immediate vicinity of the hypercritical points. Further, over the range of these experiments, the critical line is markedly asymmetric for an expansion about P_{HYP}^{L}. We require a different approach, and have developed an empirical two dimensional scaling formalism to describe scattering data in the $(T,P)_{\psi cr}$ plane. Mindful of the admonition that at no point should the relevant field variable be tangent to the critical line we have elected to develop scaling in terms of a variable, R_{red}, defined as the minimum distance from the experimental point of interest to the LL critical line in an appropriately reduced space.

$$R_{red} = \{[(t_{exp} - t_{cr})_{min}/(t_{exp} + t^{\#})]^2 + [(p_{exp} - p_{cr})_{min}/(p_{exp} + p^{\#})]^2\}^{1/2} \qquad (3)$$

In eq. 3 (t_{exp}, p_{exp}) and $(t_{cr}, p_{cr})_{min}$ are the coordinates of the experimental point of interest and that point on the critical line at minimum distance, respectively, and $t^{\#}$ and $p^{\#}$ are reduction parameters.

Further discussion is appropriate. Should the field variable lie parallel to t, an appropriate scaling equation for the correlation radius might be $\xi_t = A_t\,[|(t_{exp} - t_{cr})/(t_{cr} + t^{\#})|]^{\nu} = A_t\,\tau^{\nu}$ where τ is a reduced temperature and ν a scaling exponent. By including the constant, $t^{\#}$, difficulties are avoided, most obviously for the case when t_{cr}~0. Along the temperature coordinate that issue is often avoided by choosing t(°C) and setting $t^{\#}$=273.2, *i.e.* assuming that it is T/K which properly scales temperature. In the pressure domain we write, similarly (should the field variable lie parallel to p), $\xi_p = A_p\,[|(p_{exp} - p_{cr})/(p_{cr} + p^{\#})|]^{\nu} = A_{\pi}\,\pi^{\nu}$.

$$\xi_R = A_R\,\{[\tau^2 + \pi^2]^{1/2}\}^{\nu} = A_{\tau}\,R_{red}^{\nu}, \qquad (4)$$

i.e. after choosing a coordinate set which is orthogonalized by the transformation, the correlation radius is assumed to scale logarithmically. The reduction parameters $t^{\#}$ and $p^{\#}$ are chosen commensurately in order to insure that unit steps along $t_{red} = (t_{exp} - t_{cr})_{min}/(t_{exp} + t^{\#})$ and $p_{red} = (p_{exp} - p_{cr})_{min}/(p_{exp} + p^{\#})$ are thermodynamically equivalent. This approach avoids the canonization of any point along the critical (T,P) locus as a unique scaling reference (for

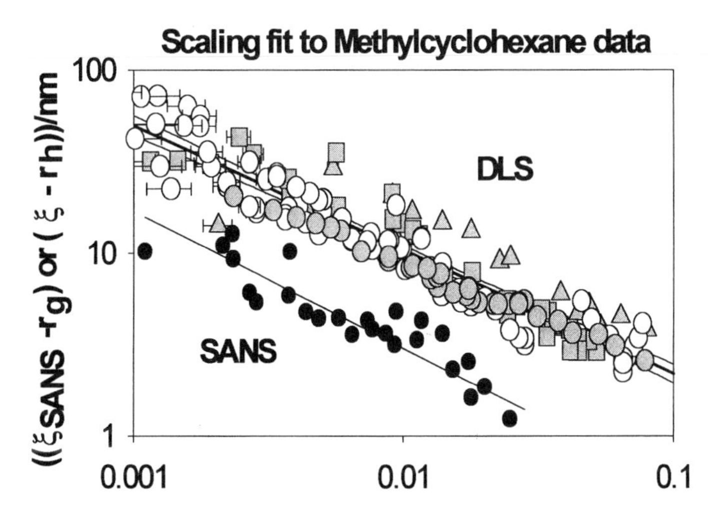

Figure 9. Logarithmic scaling representation of DLS and SANS correlation radii for demixing of PS/MCHh and PS/MCHd solutions of several M_w's including the data shown in Figure 8. Solid circles = SANS PS30k/MCHd, open circles = DLS PS30k/MCHd, shaded circles = DLS PS30k/MCHh, shaded squares = DLS PS90k/MCHh, shaded triangles = DLS PS400k/MCHh, R_{red} is defined by Equation (3) and the least squares parameters of fit reported in Table 1. Modified from ref. 22 and used with permission.

example a hypercritical temperature or pressure). As a consequence, in (ξ,R_{red}) space one avoids the doubling of critical exponents implied by such selection (see refs. 12, 22 and 23 for further discussion).

The scaling fits. R_{red} was calculated from the LL equilibrium lines shown in Figure 8, setting $t^{\#}$=273.2°C as is customary, and selecting $p^{\#}$(MCH)=55MPa by smoothing. The calculation is not sensitive to the choice of $p^{\#}$ and a universal value of $p^{\#}$ can be employed for all MCH solutions. The scaling fits to DLS and SANS, which include data at all M_w's studied is shown in Figure 9. Least squares regression lines are reported in the caption. The scaling exponents for both DLS and SANS are within experimental error of the theoretical value, ν=0.63 for three PS/solvent systems (PS/MCH, Figure 9, and PS/CH and PS/AC, not shown).[22] Dispersions of the least squares fits lie in the range 0.1 to 0.2 $\log_{10}(R_{red})$ units, and for each solvent system the DLS and SANS plots, while parallel, are offset by factors which vary from 3 (in AC) to 20 (MCH). It was this observation which led to to the development of viscosity corrections to the $(\xi_{DLS})^o$ data. The corrected DLS are now in better agreement with SANS, (*i.e.* within a factor of ~3), and in each case this holds over nearly hundred fold changes in R_{red}. It is remarkable that data in both H and D substituted solvents, and which extend over more than a decade in MW and more than two decades in R_{red}, lie on common scaling lines even though the ratio of wave lengths of the probe radiations is more than 10^3 (λ/λ_{SANS} = 1332). We conclude the two techniques are measuring identical or at closely related structural correlations.

Table 1. Least squares parameters for scaling fits of demixing (T,P) isopleths for PS/MCH/HE and PS/trans-1,4-DMCH solutions (see Equations (1) and (2)) or scaling fits of DLS and SANS correlation radii in PS/MCHh and PS/MCHd solutions (see Equations (3) and (4).

Solution		PS/MCH/HE	PS/cis/trans// 1/1-1,4DMCH	PS30k in MCHh and MCHd	
Property		Cr. Demixing Isopleth	Cr. Demixing Isopleth	DLS Correlation Radius	SANS Correlation Radius
10^{-3}(polymer M_W)		2000	575	30, 90 and 400	30
(τ_{min}, π_{min})		(273,246)	(292,249)		
	T/K	367.3	383.5		
	P/MPa	13.0	14.4		
$T_{HYP}{}^{L}$					
	T/K	340.8	348.2		
	P/MPa	119.9	163.4		
$P_{HYP}{}^{L}$					
	T/K	423.1	443.6		
	P/MPa	0.2	1.8		
A		0.735	0.750	0.460	0.096
ν		0.61	0.63	0.68	0.75
Remarks		Eq.(1) and (2) HE/MCH//20/80w/w α = 40.0 deg	Eq.(1) and (2) α = 38.3 deg	Eq.(3) and (4) t# = 273, p# =55	Eq.(3) and (4) t# = 273, p# =55

Acknowledgments: It is a pleasure to acknowledge with thanks the all important contributions of many collaborators to the research described in this paper. They include (in alphabetical order) Dr. Attila Imre (Atomic Energy Research Institute, Budapest), Dr. Marek Luszczyk (Institute of Physical Chemistry, Warsaw), Ms. Galina Melnichenko (Oak Ridge, TN), Dr. Yuri Melnichenko (Oak Ridge National Laboratory), Prof. Luis P. N. Rebelo (New University of Lisbon), Prof. Jerzy Szydlowski (University of Warsaw), and Dr. Hanna Wilczura (University of Warsaw). The research at the University of Tennessee has been supported by the U. S. Department of Energy, Division of Materials Sciences (DE88ER45374), and the Ziegler Research Fund, University of Tennessee.

References.

1. Imre, A.; Van Hook, W. A.; *Chem. Soc. Reviews* **1998**, 27, 117.
2. Imre, A.; Van Hook, W. A.; *J. Phys. Chem. Ref. Data* **1996**, 25, 637.
3. Imre, A.; Van Hook, W. A.; *Recent Res. Dev. Polymer Sci.* **1998**, 2, 539.
4. Imre, A.; Melnichenko, G.; Van Hook, W. A.; *J. Polymer Sci., Polymer Phys. Ed.* **1999**, 37, 2747.
5. Szydlowski, J.; Van Hook, W. A.; *Macromolecules* **1991**, 24, 4883.
6. Luszczyk, M.; Rebelo, L. P.; Van Hook, W. A.; *Macromolecules* **1995**, 28, 745.
7. Imre, A.; Melnichenko, G.; Van Hook, W. A.; *Phys. Chem. Chem. Phys.* **1999**, 1, 4287.
8. von Konynenburg, P. H.; Scott, R. L.: *Phil. Trans. Roy. Soc. (London)* **1980**, 298, 495.
9. Schneider, G. M.: *J. Supercrit. Fluids* **1998**, 13, 5.
10. Luna-Barcenas, G.; Meredith, J. C.; Sanchez, I. C.; Johnston, K. P.; Gromov, D. G.; de Pablo, J. J.: *J. Chem. Phys.*, **1997**, 107, 10782.
11. Saeki, S.: *Fluid Phase. Equilib.* **1997**, 136, 87.
12. Narayanan, T.; Kumar, A. *Physics Reports* **1994**, 249, 135.
13. Schneider, G. M. *Ber. Buns. Phys, Chem.* **1972**, 76, 325. *J. Chem. Thermodynamics* **1991**, 23, 301.
14. Prigogine, I.; Defay, R. *Chemical Thermodynamics*, Longmans Green, London 1954. Translated by Everett, D.H. (Chap. 17).
15. Rice, O. K. *Chem. Reviews* **1949**, 44, 65.
16. Van Hook, W. A.; Wilczura, H.; Rebelo, L. P. N.: *Macromolecules* **1999**, 32, 7299.
17. Imre, A.; Van Hook, W. A.; *J. Polymer Sci., Polymer Phys. Ed.* **1994**, 32, 2283.
18. Luszczyk, M.; Van Hook, W. A.: *Macromolecules* **1996**, 29, 6612.
19. Imre, A.; Van Hook, W. A.; *J. Polymer Sci., Polymer Phys. Ed.* **1997**, 35, 1251.
20. Scott, R. L.: J. Chem. Phys. **1949**, 17, 279..
21. Cowie, J. M. G.; McEwen, I. J.: *Polymer* **1984**, 25, 1107.
22. Van Hook, W. A.; Wilczura, H.; Imre, A.; Rebelo, L. P. N.; Melnichenko, Y.: *Macromolecules* **1999**, 32, 7312.
23. Griffiths, R. B.; Wheeler, J. C. *Phys. Rev.* **1970**, A2, 1047.

THERMODYNAMIC AND DYNAMIC PROPERTIES OF POLYMERS IN LIQUID AND SUPERCRITICAL SOLVENTS

Yuri B. Melnichenko[1], G. D. Wignall[1], W. Brown[2], E. Kiran[3], H. D. Cochran[4], S. Salaniwal[4], K. Heath[4], W. A. Van Hook[5], and M. Stamm[6]

[1]Oak Ridge National Laboratory*, Oak Ridge, TN 37831
[2]University of Uppsala, S-75121 Uppsala, Sweden
[3]University of Maine, Orono, ME 04469
[4]Oak Ridge National Laboratory and University of Tennessee, Knoxville, TN 37996
[5]University of Tennessee, Knoxville, TN 37996
[6]Institut für Polymerforschung Dresden e.V., 01069 Dresden, Germany

INTRODUCTION

Critical phenomena are grouped into classes depending on the spatial dimension (*D*) and the dimension of the order parameter (*N*). In the vicinity of the critical temperature T_C, the properties of the systems with the same (*D,N*) scale with the reduced temperature $\tau=(T-T_C)/T_C$ with identical critical indices for each property. It has long been appreciated that solutions of polymers in liquid organic solvents belong to the same universality class as simple fluids and small molecule mixtures, and their behavior near T_C is described by the critical τ-indices of the three-dimensional Ising model ($N=1$) [1]. More recently it was found that the region where polymer solutions behave universally with temperature ($\xi>>R_g$) terminates at $T=T_X$ when the correlation length ξ of the concentration fluctuations becomes equal to the radius of gyration R_g of the polymer [2,3]. At $T>T_X$ ($\xi<R_g$) the solutions enter the Θ domain where their properties are described by the mean field τ-indices.

Recent experiments have demonstrated that the Θ condition may be also induced in polymer solutions in supercritical fluids (SCFs) by varying the temperature and/or pressure [4]. A SCF is a substance at a pressure and temperature above the liquid-vapor critical point where the coexisting liquid and vapor phases become indistinguishable. The physical properties of SCFs are similar to those of dense gases, although when highly

*Managed by Lockheed Martin Energy Research Corporation under contract DE-AC05-96OR-22464 for the U. S. Department of Energy.

compressed, their density may be comparable to sub-critical liquids. Despite recent progress in many areas of SCF science and technology (see e.g. [5] and references therein), significant challenges remain in developing the same level of understanding of polymer/SCF solutions as has been reached for solutions of polymers in traditional organic solvents. This article will focus on small angle neutron scattering (SANS) and dynamic light scattering (DLS) studies of mixtures of strongly interacting (overlapping) polymers in liquid and supercritical solvents. We have observed an unexpected universality in both thermodynamic and dynamic properties of these seemingly different physical systems.

INTER- AND INTRAMOLECULAR CORRELATIONS

At the phenomenological level, liquid solvents for polymers are divided into three classes, namely poor, theta (Θ), and good solvents. The solvent quality is directly related to the ability of the solvent molecules to mediate the attractive intrachain forces responsible for the polymer – solvent demixing. Poor solvents can scarcely impede the intrachain interactions and hence can dissolve only short chain (or low molecular weight M_W) polymers with a limited number of contacts between segments. The pairwise attractive and repulsive interactions compensate at the Flory (or Θ) temperature which is defined as the upper critical solution temperature (UCST) of a polymer with infinite molecular weight $M_W = \infty$ [1]. The Θ condition corresponds to the threshold of unlimited polymer – solvent miscibility in the sense that polymer of arbitrary M_W becomes miscible in any proportion with the solvent [4]. At $T=\Theta$ the radius of gyration $R_g(\Theta)$ of polymer chains is not perturbed by excluded volume effects and is also independent of the long-range critical concentration fluctuations. In the good solvent domain $T > \Theta$ repulsive forces between segments work to expand R_g above the unperturbed dimensions at Θ. The expansion of polymer coils in semidilute solutions, i.e. solutions in which the volume fraction, ϕ, of the polymer is equal to or larger than the concentration ϕ^* at which coils begin to overlap, is expected to be smaller than in the semidilute regime due to screening of the monomer – monomer interactions (Fig.1).

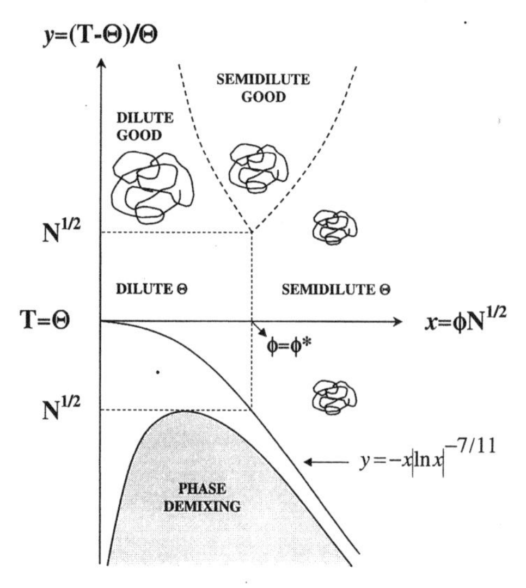

Figure 1. Generic phase diagram of polymer solutions [7,8].

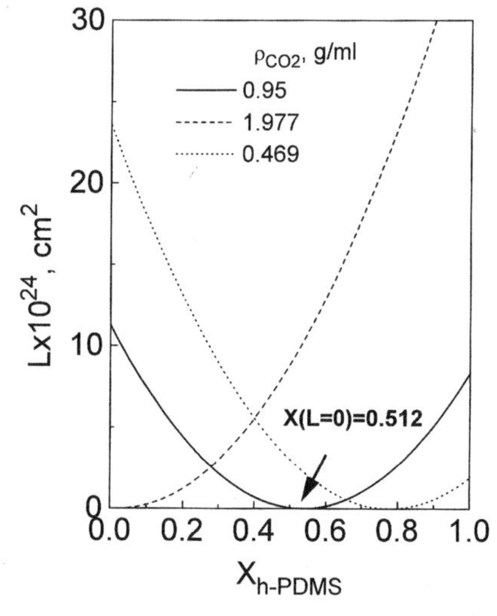

Figure 2. Variation of the parameter L in Eq.4 as a function of x, the mole fraction of PDMS-h in solutions of (PDMS-h+PDMS-d) in CO_2.

Semidilute polymer solutions are characterized by two distinct types of monomer – monomer correlations [6]. *Intramolecular* correlations between monomers which belong to the same coil are closely related to the conformation of the polymers in the solution and occur on length scales of the order of R_g. *Intermolecular* correlations are defined by fluctuations in the total concentration with a correlation length ξ. The dimension of polymer chains R_g and the correlation length ξ in polymer solutions may be determined at different thermodynamic conditions using SANS combined with the high concentration isotope labeling method (see [2] and references therein). The coherent scattering intensity, I, from an incompressible mixture of identical protonated and deuterated polymer chains dissolved in a solvent is:

$$I(Q,x) = I_s(Q,x) + I_t(Q,x) \quad , \tag{1}$$

$$I_s(Q,x) = KnN^2 S_s(Q) \quad , \tag{2}$$

$$I_t(Q,x) = LnN^2 S_t(Q) \quad . \tag{3}$$

Subscripts "s" and "t" correspond to scattering from a single chain and total scattering, respectively and thus $S_s(Q)$ is the single-chain structure factor which contains information on the intramolecular correlations. Similarly, the total scattering structure factor, $S_t(Q)$ embodies information on the total (both intra- and intermolecular) correlations between monomer units, and is related to the correlation length of the concentration fluctuations, ξ. The structure factors are normalized so that $S_s(Q=0) = 1$ and $S_t = S_s$ at infinite dilution. The scattering vector $Q = 4\pi\lambda^{-1}\sin\theta$, where 2θ is the scattering angle and λ=4.75 Å is the neutron wavelength. Also, x is the mole fraction of protonated chains relative to all (i.e. protonated and deuterated) chains on a solvent-free basis, and n and N are the number density of the polymer molecules and the degree of polymerization, respectively. The prefactors K and L are:

$$K \equiv (b_H - b_D)^2 \, x(1-x); \qquad L \equiv [b_H x + (1-x) b_D - b_S']^2 \quad , \tag{4}$$

where b_H and b_D are the scattering lengths of the protonated and deuterated monomers and b_S' is the scattering length of a solvent normalized via the ratio of the specific volumes of the monomer and the solvent molecule.

For isotopic mixtures of poly(dimethyl siloxane) (PDMS) in carbon dioxide (CO_2), b_H =0.086 10^{-12} cm, b_D = 6.33 10^{-12} cm, and b_S' =1.47 10^{-12} to 3.76 10^{-12} cm in the range between the critical (ρ_{CO2}=0.469 g/cm^3) and the liquid state densities (ρ_{CO2}=1.2 g/cm^3). For a given combination ($b_H < b_S' < b_D$), it is possible to set $L=0$ (Eq.4) at any density of SC CO_2 (Fig.2) by adjusting the concentration (x) of hydrogenated PDMS (e.g. L=0 at $x=0.512$ and ρ_{CO2}=0.95 g/cm) [9]. Similarly, for isotopic mixtures of polystyrene (PS) in deutero acetone (AC-d) (b_H =2.33 10^{-12} cm, b_D = 10.66 10^{-12} cm, and b_S' =8.88 10^{-12} cm) or PDMS in deutero bromobenzene (BrBz-d) (b_S' =5.8 10^{-12} cm) the pre-factor L becomes zero at x=0.214 and 0.085, respectively. As follows from Eqs. (1-4), the condition L=0 completely eliminates the contribution of "total scattering" and Eq.1 gives the single chain (intramolecular) scattering directly. For some solutions (e.g. PS in deutero cyclohexane, CH-d), there is no isotopic ratio, $0<x<1$, which satisfies the condition $L(x)=0$, and $I(Q,x)$ always contains a contribution from the total (intermolecular) scattering which may be subtracted via [2]:

$$I(Q,x,\tau) - \frac{L(x)}{(b_H - b'_S)^2} I_t(Q,x=1,\tau) = KnN^2 S_S(Q,\tau) \quad . \tag{5}$$

The value of the prefactor of the second term in Eq.(5) is on the order of several percent and thus the effect of this correction to $I(Q,x)$ is negligible for T >> T_C [2]. However, the correction can become finite near T_C due to the divergence of $I_t(0) \sim \tau^{-1.24}$ [10]. As soon as deuteration of the part of the polymer leads to a shift of T_C of several degrees [11], the subtraction in Eq.5 should be performed at the same values of τ rather than at the same values of the absolute temperature.

The radius of gyration R_g may be obtained by fitting S_S with the Debye function :

$$S_s^D = (2/y^2)(y - 1 + e^{-y}) \;,\; y = Q^2 R_g^2 \quad . \tag{6}$$

For solutions of non-overlapping polymer chains under Θ conditions, Eq.(6) in the limit of small Q [12,13] yields:

$$S_s(Q) = \frac{S(0)}{1 + Q^2 \xi^2(\Theta)} \tag{7}$$

and the value of the correlation length (Θ conditions) is:

$$\xi(\Theta) = R_g(\Theta)/\sqrt{3} \quad . \tag{8}$$

Thus, SANS measurements of ξ vs. temperature or pressure may be used to determine the Θ temperature or the Θ pressure using Eq.(8) [4,9]. If all chains are protonated (x=1), the prefactor K=0, and one obtains ξ(T,P) directly from $I(Q,x=1) \sim S_t(Q)$ using the Ornstein-Zernike formula:

$$S_t(Q) = S(0)/(1 + Q^2 \xi^2) \quad . \tag{9}$$

EXPERIMENTAL

Materials and Sample Preparation

Polymer samples of PS-h standards (M_W = 5700, 8300, 10200, 11600, 13000, 28000, 51500, 115000, 200000, 257900, 515000, 533000) and and PS-d standards (M_W = 10500, 11200, 119000, 205000, and 520000 of polydispersity generally $M_W/M_N \leq 1.06$ were obtained from Polymer Laboratories, USA. PS-h (M_W = 22200, 73000) was obtained from Polymer Source, Canada. PDMS-h (M_W =22500, 47700, and 79900 of polydispersity $M_W/M_N \leq 1.03$) and PDMS-d (M_W =27600, $M_W/M_N \leq 1.11$ and 49600, $M_W/M_N \leq 1.03$) were synthesized and characterized at the Max Planck Institute for Polymer Research, Germany. PDMS-d (75600, $M_W/M_N \leq 1.3$) was purchased from Polymer Standards Service GmbH, Mainz, Germany. Protonated MCH-h and BrBz-h as well as deutero substituted solvents CH-d, AC-d, MCH-d, and BrBz-d with D/(H+D)=0.995 were obtained from Sigma Chemical and dried over molecular sieves prior to preparing solutions. CO_2 (SFC purity 99.99%) was obtained from Matheson Gas products, Inc., USA.

PS – MCH and PS – CH solutions were prepared at the critical volume fraction ϕ_C of the polymer which was assigned in accordance with the empirical power law $\phi_C=6.65(M_W)^{-0.379}$ [14] and $\phi_C=7.69(M_W)^{-0.385}$ [15], respectively. PS AC-d solutions were prepared at near critical concentration C=20.3 wt%. Solutions of PDMS in SC CO_2 and in BrBz were prepared at the overlap concentration C*=0.1397 and 0.0959 g/ml for M_W 22500 and 47700, respectively calculated using $C^*=3M_W/4\pi(R_g)^3N_A$, where N_A is the Avogadro number. The overlap concentration is approximately the same as the critical concentration of a polymer solution [16]. The liquid solutions, PS-CH, PS–MCH, and PDMS-BrBz, were filtered through Millipore filters (0.22 μm) and thoroughly homogenized either in a 10 mm cylindrical optical cell (DLS) or in a 2 mm thick quartz cells (SANS) ~ 10 degrees above the Θ temperature of PS-CH (40 ^OC), PS-MCH (~ 78 ^OC) PDMS-BrBz (~ 68 ^OC) [17]. In PDMS – SC CO_2 experiments, the polymer was loaded into a stainless steel cylindrical cell having three optically polished sapphire windows one of which was located at the scattering angle $\theta=90$ ^{O} and used for the DLS measurements. The samples were pressurized with CO_2 at T=50 ^OC, stirred thoroughly for 10 min until a transparent homogeneous solutions was obtained after which the variation of the static correlation length (SANS) or the dynamic correlation length (DLS) were measured at various temperatures and/or pressures. The temperature of the samples was controlled to better than ±0.1 K. The range of T covered in the experiment was from the Θ temperature down to T_C of each solution. The critical temperature of phase demixing was identified as a sharp maximum in the integral neutron count rate as a function of T (SANS) or as the appearance of the meniscus (DLS). This allowed an estimate of T_C to better than ±0.2 K.

Small-Angle Neutron Scattering

Measurements were performed on the 30-m SANS spectrometer at the Oak Ridge National Laboratory. The neutron wavelength was $\lambda=4.75$ Å ($\Delta\lambda/\lambda=0.06$) and the range of scattering vectors was $0.005<Q=4\pi\lambda^{-1}\sin\theta<0.05$ Å^{-1}, where 2θ is the scattering angle. The data were corrected for scattering from the empty cells, detector sensitivity and beam-blocked background and placed on an absolute scale using pre-calibrated secondary standards after radial averaging to produce functions of the intensity *I vs. Q*. Procedures for subtracting the incoherent background have been described previously [18]. The functions $S_S(Q,T)$ were obtained using Eqs.1,2 (PS-AC-d, PDMS-BrBz-d, PDMS-SC CO_2) and Eq.5 (PS-CH-d) and used to extract the (z-averaged) $R_g(T)$ by fitting the Debye form factor Eq.6. The functions $S_t(Q,T)$ were obtained using Eqs.1,3 and used to extract the correlation length ξ(T) using Eq.9.

Dynamic Light Scattering

DLS measurements were made in the self-beating (homodyne) mode using a setup the details of which were described previously [19]. The cylindrical scattering cells with liquid polymer solutions were sealed and immersed in a large-diameter thermostated bath containing decalin placed at the axis of the goniometer. Measurements were made at different angles in the range 30 to 135° as a function of temperature ($80 < T/^OC < 0$). Analysis of the data was performed by fitting the experimentally measured $g_2(t)$, the normalized intensity autocorrelation function, which is related to the electrical field correlation function, $g_1(t)$ by the Siegert relation [20]:

$$g_2(t) - 1 = \beta \,|\, g_1(t) \,|^2 \quad , \tag{10}$$

where β is a factor accounting for deviation from ideal correlation. For polydisperse samples, $g_1(t)$ can be written as the inverse Laplace transform (ILT) of the distribution of relaxation times, $\tau A(\tau)$:

$$g_1(t) = \int \tau A(\tau) \exp(-t/\tau) d \ln \tau \quad , \tag{11}$$

where t is the lag-time. The relaxation time distribution, $\tau A(\tau)$, is obtained by performing the inverse Laplace transform (ILT) with the aid of a constrained regularization algorithm (REPES). The mean diffusion coefficient (D) is calculated from the second moments of the peaks as $D_C = (1-\phi)^{-2}\, \Gamma/Q^2$, where $Q = (4\pi n_o/\lambda)\sin(\theta/2)$ is the magnitude of the scattering vector and $\Gamma = 1/\tau$ is the relaxation rate. Here Θ is the scattering angle, n_o the refractive index of pure solvent and λ the wavelength of the incident light.

The Stokes-Einstein equation relates the cooperative diffusion coefficient (D_c) to the "bare" dynamic correlation length (ξ_d*) defined in terms of the temperature-dependent viscosity of the solvent (η_s) in Eq.12:

$$\xi_d^* = kT/(6\pi\eta_s D_c) \quad , \tag{12}$$

where $k_B T$ is the thermal energy factor. Another parameter of choice that may be used in the Stokes-Einstein Eq.12 is the macroscopic shear viscosity η_m. We define the "true" dynamic correlation length (ξ_d) as:

$$\xi_d = kT/(6\pi\eta_m D_c) \quad . \tag{13}$$

Viscometry

The viscosity of the blank solvents (MCH and BrBz) and polymer solutions was measured over the temperature range $0 \leq T/^{o}C \leq 80$ using a Ubbelohde-type viscometer which was sealed immediately after loading the liquids. The data for the blank solvents were fit using the second order polynomial function:

$$\eta_s = A + B(T_k) + C(T_k)^2 \tag{14}$$

where η_S is in centipoises and the temperature T_k in Kelvins. The parameters thus determined are: (A=9.3±0.9, B=-0.049±0.006, and C=(6.8±0.9)10^{-5} for MCH), and are: (A=13±0.8, B=-0.068±0.005, and C=(9.2±0.8)10^{-5} for BrBz). The viscosity of SC CO_2 was calculated using an empirical relation given by Sovova and Prochaska [22] (200 < T/K < 1500, 0 < P/MPa < 100):

$$\eta_S = [(18.56 + 0.014T_k)(\rho^{-1} - 7.41\text{x}10^{-4} + 3.3\text{x}10^{-7} T_k)]^{-1}$$

where η_S is the viscosity of SC CO_2 in [Pa s] and ρ is the density of SC CO_2 in kg m^3. The viscosity of PS-MCH and PDMS-BrBz solutions was fit to a function obtained from mode coupling theory [23]:

$$\eta_m/\eta_b = B\tau^{-z\nu} \quad , \quad \eta_b = \eta_o \exp(A/T_k) \quad . \tag{15}$$

The product $z\nu$ is predicted to be 0.034 and B is a constant of the order of unity. The"background" viscosity η_b of PS-MCH solutions measured far from the critical demixing temperatures is well described with the values of the parameters η_O = (2.2 ± 0.3) 10^{-3} cP, and A = 2700 ± 400 over the range (5700 ≤ M_W ≤ 115000).

RESULTS AND DISCUSSION

Structure and Thermodynamic Properties at $\Theta \geq T \geq T_C$

The different theoretical approaches that have been applied to calculate R_g at the critical demixing point (T^c,ϕ^c) have led to conflicting conclusions. Starting from de Gennes' assumption that *at the critical point polymer chains just begin to overlap and do not interpenetrate significantly* [16] it was shown that the polymer chains should be partially collapsed at $T \cong T_C$ and $\phi = \phi_C$ [24]. Conversely, the molecular theory [6,25] and computer simulations [26] show that conformation of individual chains below the Θ temperature should remain unperturbed and collapse to the globular state should occur only at $T<T_C$. The temperature variation of the dimension of polymer chains at $\Theta \geq T \geq T_C$ was first explored using SANS along with the high concentration isotope labeling in [2] for solutions of PS in the Θ solvent CH-d. As is seen from Fig.3, R_g of interacting PS coils does not decrease over the whole temperature range below the Θ temperature. At the same time, the correlation length of the concentration fluctuations ξ which is much smaller than R_g at T~Θ diverges in the vicinity of T_C. These findings indicate that the deterioration of the quality of Θ solvents leads to the formation of microdomains of the size ξ, representing clusters of unperturbed, strongly interpenetrating polymer coils.

The region $\xi > R_g$ is realized in the vicinity of the critical demixing temperature and is thus associated with the poor solvent domain. The solvent quality can be usually improved by increasing the temperature or pressure, to move the solution away from the phase boundary. In Θ solvents (e.g. PS – CH) the solution is capable of reaching the Θ condition at some T and P whereas in poor solvents, e.g. PS – AC the condition cannot be reached at any accessible T, P. The effect of pressure on the thermodynamic state of PS/AC-d solutions is illustrated in Fig.4. Within experimental error, R_g remains independent of pressure down to the critical demixing P_C. Although the Θ condition does not exist for PS-AC solutions, the values for R_g remain close to the unperturbed dimensions for Gaussian chains with M_W=11600 (i.e. $R_g \cong 0.27\ M_w^{1/2} \sim 29$ Å). The correlation length, ξ, diverges as P falls to P~P_C, where the system exhibits pressure-induced phase demixing, but for solutions of PS in AC, the magnitude of ξ remains generally ≥ R_g [9]. This delineates an important difference between the structure of the liquid Θ and poor solvents. The size ξ of the concentration fluctuations in polymer solutions in Θ solvents can decrease below the dimension of the constituent chains and reach the value of ξ(Θ) defined by Eq.8 (Figs.3, 5) whereas the value of ξ in poor solvents always remains > ξ(Θ) (Figs.4, 6) and levels off at some thermodynamic distance from the critical point.

Supercritical fluids in general and supercritical carbon dioxide in particular have emerged as an attractive alternative to the organic solvents used for polymer manufacturing and processing. One key advantage of SCFs is the possibility of continuously tuning the solvent quality by varying the pressure (or density) in addition to the temperature. Due to the high compressibility of SCF solvents the density may be

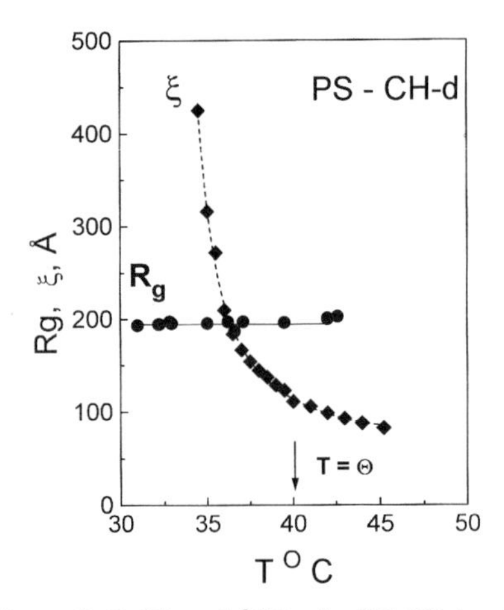

Figure 3. $R_g(T)$ and $\xi(T)$ for PS-CH-d solution (M_W=533000) at ambient conditions. R_g measured over the range of pressure $0.1 \leq P \leq 50$ MPa [9].

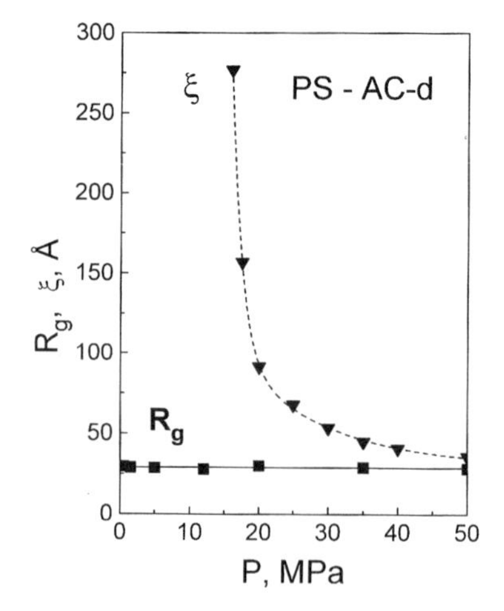

Figure 4. $R_g(P)$ and $\xi(P)$ for PS-AC-d solution (M_W=11600) at T=30 °C [9].

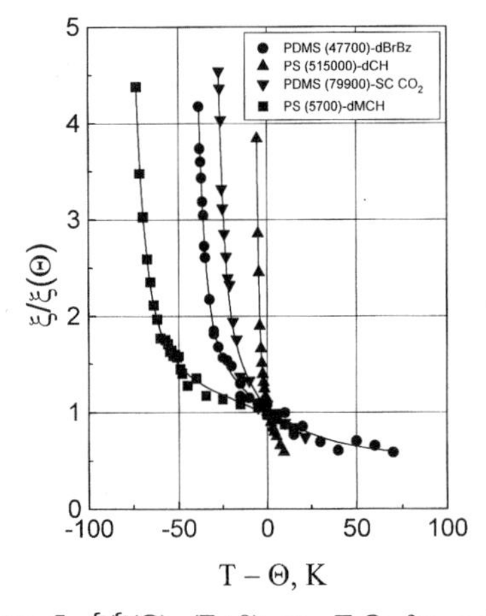

Figure 5. $\xi/\xi(\Theta)$ (Eq.8) vs. T-Θ for polymer solutions in Θ solvents specified in the inset.

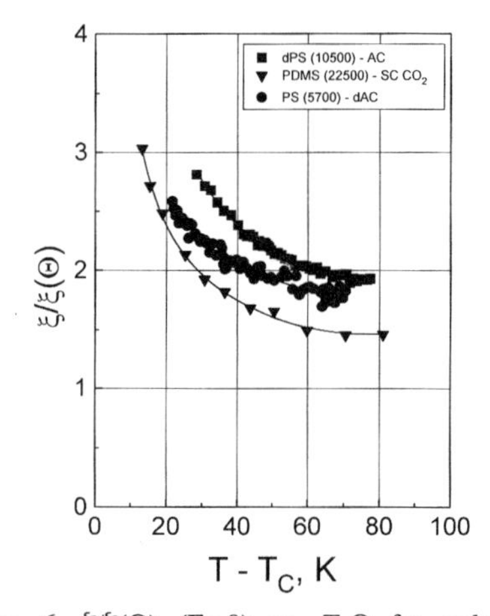

Figure 6. $\xi/\xi(\Theta)$ (Eq.8) vs. T-Θ for polymer solutions in poor solvents specified in the inset.

conveniently varied with pressure over a much wider range (factors of 2 to 10) than is possible for organic liquid solvents where the range is at most a few percent. This property of SCF solvents should offer an unparalleled means of controlling polymer solubility, but unfortunately, only two classes of polymeric materials, i.e. fluoropolymers and silicones have been shown to exhibit appreciable solubility in SC CO_2 [27]. The first SANS measurements of the second virial coefficient A_2 were performed in mixtures of hexafluoropropylene oxide (HFPPO) and poly(1,1-dihydroperfluoro-octylacrylane) (PFOA) with SC CO_2 at T=65 °C and P=34.5 MPa (see [28] and references therein). They have shown that $A_2 \cong 0$ for the former and $A_2 > 0$ for the latter systems and thus SC CO_2 is a Θ solvent and a good solvent for HFPPO and PFOA, respectively at these particular values of (T,P). In a recent work [4] SANS was applied to investigate the structure and thermodynamic properties of PDMS – SC CO_2 solutions as a function of pressure and temperature. Fig.7 illustrates the (T,P) dependence of the radius of gyration of PDMS in SC CO_2. R_g of the polymer remains invariant during both pressure and temperature quenches extending to the immediate vicinity of the (T,P) polymer-SCF demixing locus and agrees well with the dimension of unperturbed chains

($R_g = 0.267M_w^{1/2} \cong 40$ Å). This observation indicates the universality of the constancy of R_g in semidilute polymer solutions below the Θ condition previously demonstrated for liquid solvents (see Figs.3, 4) and now extended to supercritical fluids. The dimensions of polymer chains increase when T and/or P exceed the appropriate values of the Θ parameters ($\Theta = 65 \pm 5$ ^{o}C and $P_\Theta = 52 \pm 4$ MPa at ρ_{CO2}=0.95 g/cm^3 [4]) due to excluded volume effects, indicating that SC CO_2 becomes a good solvent for PDMS. The experimental data agree with the results of Monte Carlo simulations which suggest that polymer chains may adopt unperturbed and expanded conformations at high densities [29].

A typical variation of ξ vs. τ in PDMS-SC CO_2 solutions is shown in Fig.8. As is seen, the critical index ν exhibits a sharp crossover from the mean-field value (ν=0.5) in the Θ region to the Ising model value (ν=0.630$\pm$0.001 [30]) in the region around T_C. Accordingly, the critical index γ=1.23$\pm$0.02 for the susceptibility (i.e. the osmotic compressibility) agrees within experimental errors with the Ising model value (1.239$\pm$0.002 [30]). The crossover takes place at $\xi \sim R_g$ and thus reproduces the main features of the crossover observed in solutions of PS in CH-d [3]. The dimension of the poor solvent domain may be made infinitely large (as in PS-AC) or small (as in PS-CH) depending on the density of SC CO_2. These observations certainly establish basic similarities of behavior in solutions of polymers in SCFs and in sub-critical solvents, and show that SCF/polymer solutions belong to the Ising model universality class.

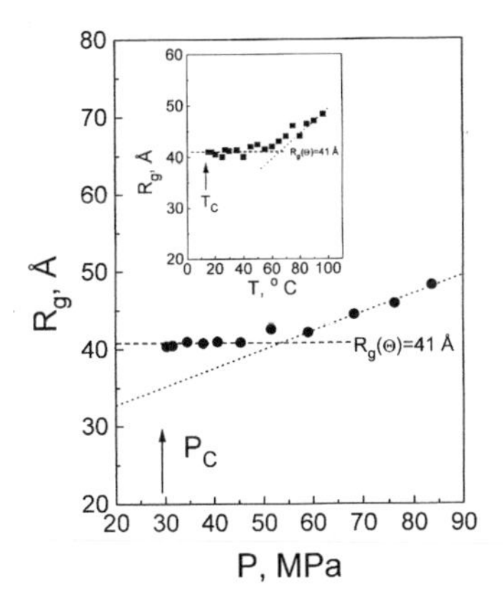

Figure 7. R_g (P) of PDMS (M_W=22500) in SC CO_2 at T=70 ^{o}C. The inset shows R_g (T) at ρ_{CO2}=0.95 g/cm^3.

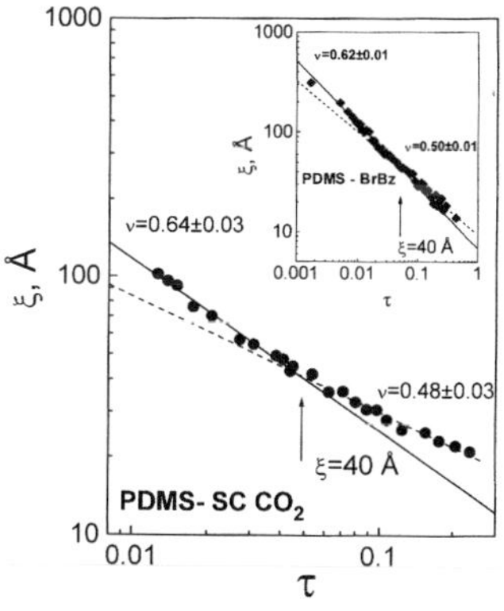

Figure 8. $\xi(\tau)$ of PDMS (M_W=22500) in SC CO_2. ρ_{CO2}=0.95 g/cm^3. The inset shows $\xi(\tau)$ for the same polymer in BrBz-d.

Dynamic Properties at $\Theta \geq T \geq T_C$

Despite numerous efforts, the dynamic properties of semidilute polymer solutions in organic solvents at $T \leq \Theta$ are poorly understood. To our knowledge, only one DLS experiment has been performed so far in the poor solvent domain of supercritical polymer solutions [4]. The most important issue which is yet to be resolved is whether the viscosity of the solvent or that of the solution should be used to calculate the dynamic correlation length from the Stokes-Einstein Eq.12 [31].

In terms of a traditional description of critical phenomena, the temperature variation of the dynamic and static correlation lengths may be represented as [32]:

$$\xi_d^* = \xi_0^* |\tau|^{-\nu^*} \quad , \tag{16}$$

$$\xi = \xi_0 |\tau|^{-\nu} \quad . \tag{17}$$

A master plot for ξ_d* as a function of τ is shown in Fig.9. As is seen, the value of the critical exponent ν* in Eq.16 for all solutions studied in the hydrodynamic regime $q\xi<<1$ around the Θ temperature is ν*=0.8±0.1 which value as yet has no theoretical justification. At smaller τ, generally at $\tau \sim 0.01$, the decay rate becomes independent of temperature characterizing entry to the critical non-diffusive regime [33]. There is a consequent observed "flattening-out" of the bare dynamic correlation length, defined via Eq.12, as a function of τ to a limiting value as T_C is approached. The variation of the dynamic correlation length with τ changes dramatically if Eq.(13) is used to define ξ_d due to significant difference in the temperature variation of η_s and η_m described by Eq. 14 and 15, respectively (Fig10). In the vicinity of the Θ temperature the temperature variation of ξ_d and similarly that of the static correlation length, is described by the mean field critical index ν=0.5 (Eq.17). At some "intermediate dynamic crossover temperature" T_X* the index increases abruptly and the crossover to the critical non-diffusive regime is observed. We note that in the "intermediate" region of τ the critical index ν=0.72 exceeds the theoretical value of ν=0.63 probably due to a complex interference between the dynamic and static correlations of the order of $\xi(\Theta)$ and/or R_g (see Eqs.7 and 8). At the same time, the crossover to the critical non-diffusive regime occurs at about the same value of $\tau \sim 0.01$ for both ξ_d and ξ_d*.

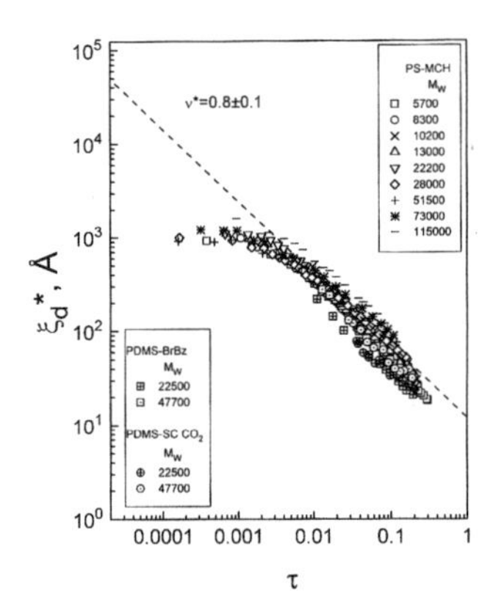

Figure 9. The bare dynamic correlation length (Eq.12) vs. τ. The symbols are spelled out in the insets.

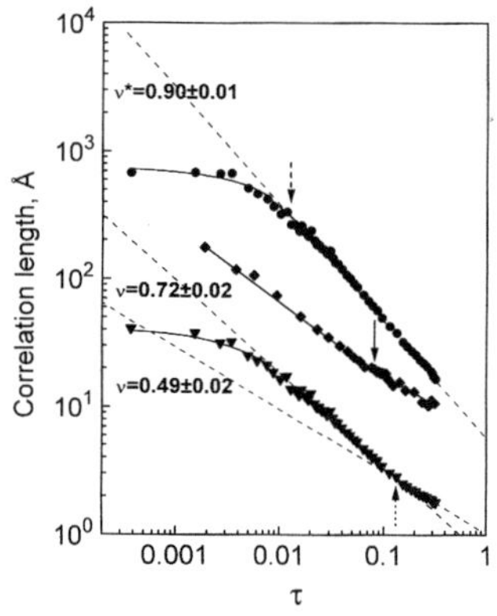

Figure 10. ξ_d* , ξ_d, and ξ_s for solution PS (M_W=5700) in CH-d. The solid and dashed arrows show the static and dynamic crossovers.

The description of the temperature variation of the correlation lengths (Eqs.16, 17) does not take into account the existence of the tricritical Θ temperature in the phase diagram of polymer solutions (Fig.1). A more general scaling description may be achieved by using the scaling variable [16]:

$$\tau' = (T - T_C)/(\Theta - T_C) \tag{18}$$

which accounts for the temperature distance from both the Θ temperature and T_c. In terms of τ', Eq.17 may be represented as [12]:

$$\xi = \xi\,(\Theta)(\tau')^{\nu} \quad , \tag{19}$$

where the amplitude $\xi(\Theta)$ is given by Eq.(8). In the case of the dynamic correlation length no theoretical interpretation for $\xi_d*(\Theta)$ has yet been developed. However, one may suggest phenomenologically that ξ_d* should follow the same scaling as Eq.19, writing

$$\xi_d^* = \xi_d^*(\Theta)(\tau')^{\nu *} \quad , \tag{20}$$

where $\xi_d*(\Theta)$ is the experimental value of the dynamic correlation length at T=Θ (Fig.11). As is seen in Fig.12, the reduced correlation lengths $\xi_d*/\xi_d*(\Theta)$ for various polymers in different solvents fall on the same master curve when plotted vs. the reduced temperature τ'. This is the first demonstration of the universality for the dynamic behavior of solutions of polymers in organic liquids and supercritical fluids.

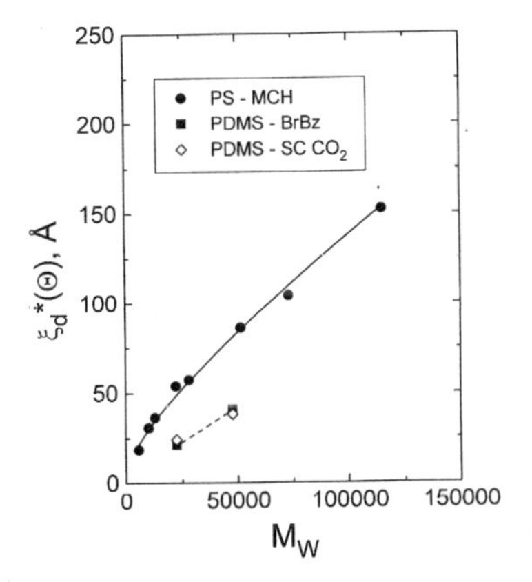

Figure 11. $\xi_d*(\Theta)$ for several polymer solutions. The symbols are spelled out in the insets.

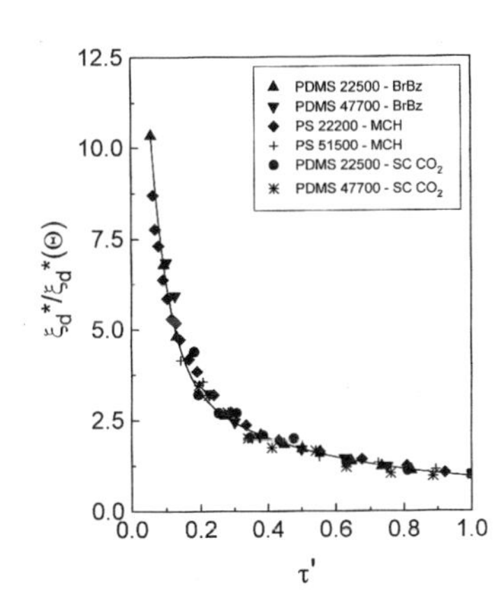

Figure 12. Master plot for $\xi_d/\xi_d*(\Theta)$ vs. de Gennes' scaling variable (Eq.18).

CONCLUDING REMARKS

Small angle x-ray scattering and static light scattering have been applied since the 1940s to obtain structural information on polymer solutions, though the limit of zero concentration was required to eliminate interchain interference. Thus, these methods could not be applied to the solutions of interacting macromolecules, due to the difficulties of separating the *inter*- and *intra*-chain contributions to the structure. However, SANS has removed this limitation and due to a fortuitous combination of several factors: high bulk penetrating power, the ability to manipulate local scattering amplitudes through isotopic labeling or an appropriate choice of solvent (contrast variation), the technique has developed into an extremely powerful tool for the study of polymers. By deuterating a fraction of the polymer, it is possible to measure the single-chain structure factor and thus the R_g of the individual polymer chains in strongly correlated systems. That approach provides exceptional means for investigating the structure of various complex fluids, e.g. solutions of synthetic and biological polyelectrolytes. Recent SANS and DLS measurements demonstrate a striking universality between structure, thermodynamic and dynamic properties of sub- and supercritical polymer solutions.

ACKNOWLEDGEMENTS

The research was supported the Divisions of Advanced Energy Projects, Materials Sciences, and Chemical Sciences, U. S. Department of Energy under contract No. DE-AC05-96OR-22464 (with Lockheed Martin Energy Research Corporation) and Grant No. DE-FG05-88ER-45374 (with the University of Tennessee) and by the National Science Foundation Grant No. CTS-9613555 (with the University of Tennessee). The financial support of YBM by Alexander von Humboldt-Stiftung is gratefully acknowledged.

REFERENCES

1. I. C. Sanchez. *Encyclopedia of Physical Science and Technology*, Academic Press, New York (1992).
2. Y. B. Melnichenko and G. D. Wignall, Dimension of polymer chains in critical semidilute solutions, *Phys. Rev. Lett.* 78: 686 (1997).
3. Y. B. Melnichenko, M. A. Anisimov, A. A. Povodyrev, G. D. Wignall, J. V. Sengers, and W. A. Van Hook, Sharp crossover of the susceptibility in polymer solutions near critical demixing point, *Phys. Rev. Lett.* 79: 5266 (1997).
4. Y. B. Melnichenko, E. Kiran, G. D. Wignall, K. D. Heath, S. Salaniwal, H. D. Cochran, and M. Stamm, Pressure- and temperature-induced transitions in solutions of poly(dimethyl siloxane) in supercritical carbon dioxide, *Macromolecules* 32: 5344 (1999).
5. M. A. McHugh and V. J. Krukonis. *Supercritical fluid extraction: principles and practice*, Butterworth-Heinemann, Boston (1994).
6. M. Muthukumar, Thermodynamics of polymer solutions, *J. Chem. Phys.* 85: 4722 (1986).
7. M. Daoud and G. Jannink, Temperature-concentration diagram of polymer solutions, *J. Phys. (Paris)* 37: 973 (1976).
8. B. Duplantier, Lagrangian tricritical theory of polymer chain solutions near the Θ-point, *J. Phys. (Paris)* 43: 991 (1982).
9. Y. B. Melnichenko,G. D. Wignall, W. A. Van Hook, J. Szydlowsky, H. Wilczura, and L. P. Rebelo, Comparison of inter- and intramolecular correlations of polystyrene in poor and Θ solvents via small-angle neutron scattering, *Macromolecules* 31: 8436 (1998).
10. Y. Melnichenko, M. Agamalyan, V. Alexeev, V. Klepko, and V. Shilov, Crossover between tricritical and critical demixing regimes of polymer solutions. Small-angle neutron scattering, *Europhys.Lett.* 19: 355 (1992).
11. C. Strazielle and H. Benoit, Some thermodynamic properties of polymer-solvent systems. Comparison between deuterated and undeuterated systems, *Macromolecules* 8: 203 (1975).
12. J. Des Cloizeaux and G. Jannink, *Polymers in Solution. Their Modeling and Structure*, Claredon Press, Oxford (1990).
13. H. Fujita, *Polymer Solutions*, Elsevier, Amsterdam, (1990).
14. T. Dobashi, M. Nakata, and M. Kaneko, Coexistence curve of polystyrene in methylcyclohexane, 1. Range of simple scaling and critical exponents, *J. Chem. Phys.* 72: 6685 (1980).
15. R. Perzynski, M. Delsanti, and M. Adam, Experimental study of polymer interactions in a bad solvent, *J. de Physique* 48: 115 (1987).
16. De Gennes P.-G., *Scaling Concepts in Polymer Physics*, 2nd ed., Cornell University Press, Ithaca (1979).
17. *Polymer Handbook*, J. Brandrup, E. H. Immergut, E. A. Grulke, eds, John Wiley, New York (1999).
18. G. D. Wignall and F. S. Bates, Absolute calibration of small-angle neutron scattering data, *J. Appl. Crystallogr.* 20: 28 (1987).
19. Y. B. Melnichenko, W. Brown, S. Rangelov, G. D. Wignall, and M. Stamm, *Phys. Lett.*, in press.
20. B. Chu, *Laser Light Scattering*, Academic Press, New York (1991).
21. W. Brown and T. Nicolai, Scattering from concentrated polymer solutions, in: *Dynamic Light Scattering*, W. Brown, ed., Clarendon Press, Oxford (1993).
22. H. Sovova and J. Prochaska, Calculations of compressed carbon dioxide viscosities, *Ind. Eng. Chem. Res.* 32: 3162 (1993).
23. H. C. Burstyn and J. V. Sengers, Decay-rate of critical concentration fluctuations in a binary liquid, *Phys. Rev.* A25: 448 (1982).

24. Y. Izumi and Y. Miyake, Universality of the coexistence curves in a polymer solutions, *J. Chem. Phys.* 81: 1501 (1984).
25. G. Raos and G. Allegra, Chain collapse and phase separation in poor-solvent polymer solutions: a unified molecular description, *J. Chem. Phys.* 104: 1626 (1996).
26. N. B. Wilding, M. Muller, and K. Binder, Chain length dependence of the polymer-solvent critical point parameters, *J.Chem.Phys.* 105: 802 (1996).
27. C. F. Kirby and M. A. McHugh, Phase behavior of polymers in supercritical fluid solvents, *Chemical Reviews*, 99: 565 (1999).
28. G. D. Wignall., Neutron scattering studies of polymers in supercritical carbon dioxide, *J. Phys.: Condens. Matter* 11: R157 (1999).
29. G. Luna-Barcenas, J. C. Meredith, I. C. Sanchez, K. P. Johnson, D. G. Gromov, and J. J. de Pablo, Relashionship between polymer chain conformation and phase boundaries in a supercritical fluid, *J. Chem. Phys.* 107: 10782 (1997).
30. J. C. Le Guillou and J. Zinn-Justin, Critical exponents from field theory, *Phys. Rev.* B21: 3976 (1980).
31. S. F. Edwards, private communication.
32. N. Kuwahara, D. V. Fenby, M. Tamsky, and B. Chu, Intensity and line width studies of the system polystyrene-cyclohexane in critical region, *J. Chem. Phys.* 55: 1140 (1971).
33. P. Stepanek, Critical dynamics of binary liquid mixtures and simple fluids studied using light scattering, in: *Dynamic Light Scattering*, W. Brown, ed., Clarendon Press, Oxford (1993).

THE COHESIVE ENERGY DENSITY OF POLYMER LIQUIDS

G. T. Dee and Bryan B. Sauer

DuPont Experimental Station
Wilmington, DE, USA

Introduction

One of the most important thermodynamic properties of a liquid is its cohesive energy density (CED). It is the energy per unit volume of the liquid required to move a molecule from the liquid state to a vapor state. It is a measure of the interactions between the molecules in the liquid state. The CED is defined as the ratio of the molar energy of vaporization (ΔU) divided by the molar liquid volume (V). Polymer molecules have a vanishingly small vapor pressure and we cannot compute an energy of vaporization. This has led to a number of approximate methods which are currently used to compute the CED of polymer liquids. Many of these are empirical and involve group contribution methods[1] based on small molecule liquids. Other methods employ pressure-volume-temperature (PVT) data to approximate the CED based on its relationship to the internal pressure ($\Pi = (\partial U/\partial V)|_T$).[1,2] This later method is commonly used in equation of state calculations for polymer systems, but the validity of the relationship between the internal pressure and the CED is rarely discussed and is sometimes misunderstood or ignored. A limited experimental method used to estimate the CED of a polymer liquid is to study the "swelling" of the polymer by the liquid of known CED.[3] This approach is not practical for a detailed determination of the properties of the CED for polymer liquids, however.

The solubility parameter (δ) is defined as the square root of the CED. This quantity was indroduced by Hildebrand[2] and used as a predictor of liquid-liquid solubility. The enthalpy of mixing is proportional to square of the difference (δ_1-δ_2) in the solubility parameters for the two liquids. Thus in combination with the entropy of mixing, the magnitude of this quantity determines the free energy of mixing. For polymers, where we do not have an experimental method for the determination of the CED of the liquid, we cannot expect to be able to make accurate predictions for the enthalpy of mixing. This point is elaborated further below.

In this paper we attempt to construct a scaling or corresponding states methology for the prediction of the CED using a combination of the surface tension and PVT

Computational Studies, Nanotechnology, and Solution Thermodynamics of Polymer Systems
Edited by Dadmun *et al.*, Kluwer Academic/Plenum Publishers, New York, 2000

measurements for polymer liquids. It has long been understood that the surface tension is a direct manifestation of the cohesive forces that hold liquids together. Correlations between the CED and the surface tensions of small molecule liquids have been demonstrated.[3] However, in the case of polymer liquids, accurate surface tension data have not been available until recently.[4] The measurment of the thermodynamic properties of polymer liquids is difficult due to the fact that the temperature domain of the liquid state is frequently high enough for degradation reactions to occur which detract from a accurate measurement of many thermodynamic properties. PVT data for polymer liquids have also only recently become available due to the need for special dilatometers for the measurement of viscoelastic liquids.[5] The above mentioned stability problem necessitates a rapid measurement of the surface tension. This rules out conventional methods due to the long relaxation times associated with these polymer liquids. Sauer[6] developed a new method which allows for the rapid determination of the surface tension of a highly viscous liquid using a very fine glass fiber.

The Internal Pressure

It was Hildebrand that first made the observation that the internal pressure (Π) was approximately equal to the CED for a number of hydrocarbon liquids.[2] The internal pressure is defined by the following equation:

$$\Pi = (\partial U/\partial V)|_T = T\,(\alpha/\beta) - P \;\approx T(\alpha/\beta) \text{ (for P<1MPa)} \tag{1}$$

where α is the thermal expansivity, β is the isothermal compressibility, T is the temperature, and P is the pressure. Many authors have collected data at room temperature which supports this approximate observation.[7] The ratio of the internal pressure to the CED was found to be close to 1 for many hydrocarbon liquids at room temperature. Hildebrand adopted the notation n ($n = \Pi/\text{CED}$) for this ratio.[2] For polar and hydrogen bonded liquids, the value of n deviated substantially from 1 and usually took on values less than 1. Weakly bonded liquids such as the perfluoro alkanes exhibited values of n greater than 1. There is no physical reason why the value of n should be equal to 1 except for the special case of a van der Waals fluid.[3] The assumption that the internal pressure is equal to the CED is subject to substantial errors depending on the chemical structure of the molecules making up the liquid.

Many equation of state theories make the assumption that the Π is equal to the CED. It is implicit in the theories of Flory, Orwoll, and Vrij,[8] (FOV) and in the Sanchez Lacombe.[9] Therefore, when one uses PVT data to define the parameters in these models one is using the internal pressure as a measure of the CED of that liquid. Any quantity computed in this manner will be subject to the error mentioned in the previous paragraph. Since Π is related to the CED by a simple multiplicative constant, one might assume that this is not a major problem and in many cases this shortcoming simply disappears into an adjustable parameter. This is not the case when one is dealing with mixtures. If one computes the enthalpy of mixing with one of the above mentioned equations of state, one arrives at a term with the following form if we assume ideal random mixing:

$$\Delta H \sim (\delta_1 - \delta_2)^2 \tag{2}$$

If we use the internal pressures of the two components to compute this term we obtain the following expression:

$$\Delta H \sim (\sqrt{\Pi_1} - \sqrt{\Pi_2})^2 = [\sqrt{(n_1.CED_1)} - \sqrt{(n_2.CED_2)}]^2$$

$$= (n_1) [[\sqrt{(CED_1)} - \sqrt{(n_2/n_1)} (\sqrt{(CED_2)}]^2 \qquad (3)$$

Equation 3 shows that we do not have a good measure of the enthalpy of mixing, nor is the difference simply offset by a multiplicative constant. The problem is further compounded by the fact that the ratios n_i may be functions of the thermodynamic variables of the system i.e. $n_i = n_i (T,..)$. If the values of n_i are similar and have values close to one then the enthalpy of mixing can be approximated. A second possibility is the case where the liquids have similar values of n_i but n_i is different from unity. In this case the enthalpy of mixing will be in error by a multiplicative constant and can be corrected by a fudge factor, as is the common practice in this field.

The Internal Pressure and Surface Tension.

The surface tension of liquids, γ, is a direct manifestation of the cohesive forces that hold the liquid state together. We have been working in the area of the surface tension of polymer liquids. We have acquired a substantial amount of data for high molecular weight liquids.[10] PVT properties were also obtained for these same polymer samples.[11] By scaling the surface tension data with an appropriate measure of the CED of the liquid, we should be able to collapse the data for many different polymers onto a single master curve. Implicit in this assumption is the fact that the CED and the density of the liquid are the dominant factors contributing to the surface tension. We can scale the surface tension using Π computed from the PVT data and the measured density. Patterson and Rastogi[12] were the first to do this for polymer liquids. Clearly, any quantity derived from this information set (γ, v_{sp}, α, β,T, P) can be used to scale the data, where v_{sp} is the specific volume, α is the thermal expansivity, β is the isothermal compressibility, T is the temperature , and P is the pressure. In particular we can use the equation of state parameters obtained from fitting an equation of state to the PVT data.[13] The master curve will be different but the information content will be the same. Figure 1 shows this exercise for a number of polyethylenes ranging from low molecular weight alkanes to rather large molecules. The striking feature about the PE data set is the complete collapse of the data to within a 2% scatter. The measured surface tensions for these liquids vary by over 200 % in this molecular weight range.[10] Since the chemical structure of these liquids is similar at the local molecular level, we expect that the values of n for these molecules will be close in value and the complete collapse of the data onto a single curve reflects the dominant role played by the CED in the determination of the surface tension.

We repeated this exercise for polymer molecules of varying molecular weights but with different chemical structures.[13] Figure 1 shows the result of the scaling for poly(ethylene oxide) (PEO), Poly(dimethyl siloxane) (PDMS), polystyrene (PS), and polyethylene (PE). For molecules with the same chemical structure but different molecular weights we observe a similar collapse of the data for each polymer type. However, we now observe that we have a different master curves for each polymer type. The obvious explanation is again to observe that we are not scaling with the true CED of these liquids and that in each case there is a shift factor coming from the function n_i for each chemical structure. It is tempting to assume that if we knew the values of n_i for each of these different polymer families, we could achieve a complete collapse of the data set onto a single master curve. The central idea of this paper is to turn this last idea around and propose that such a master curve exists. If we can determine this curve for one

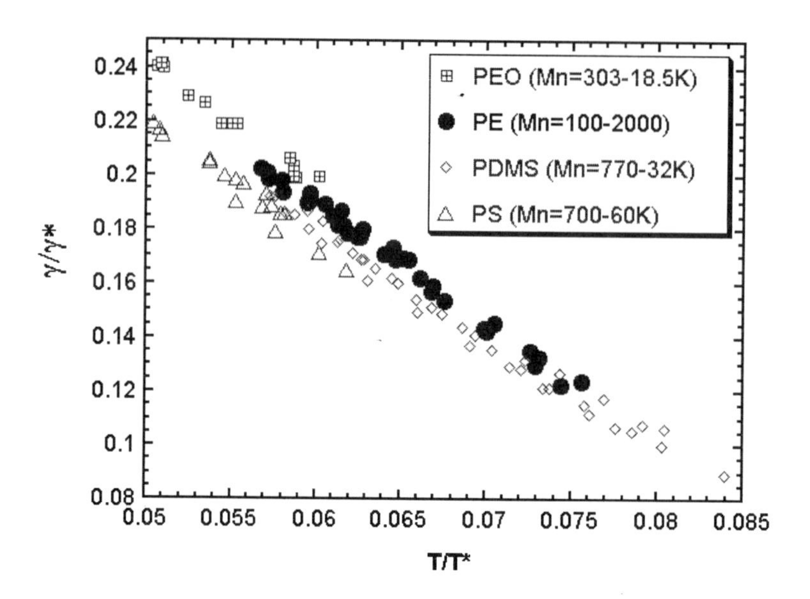

Figure 1: A plot of the scaled surface tension, γ/γ^*, as a function of the scaled temperature T/T^*. $\gamma^* = P^{*1/3} T^{*2/3} (ck)^{1/3}$ where P^* and T^* are temperature dependant FOV equation of state fitting parameters, k is Boltzmann's constant and c is a constant.

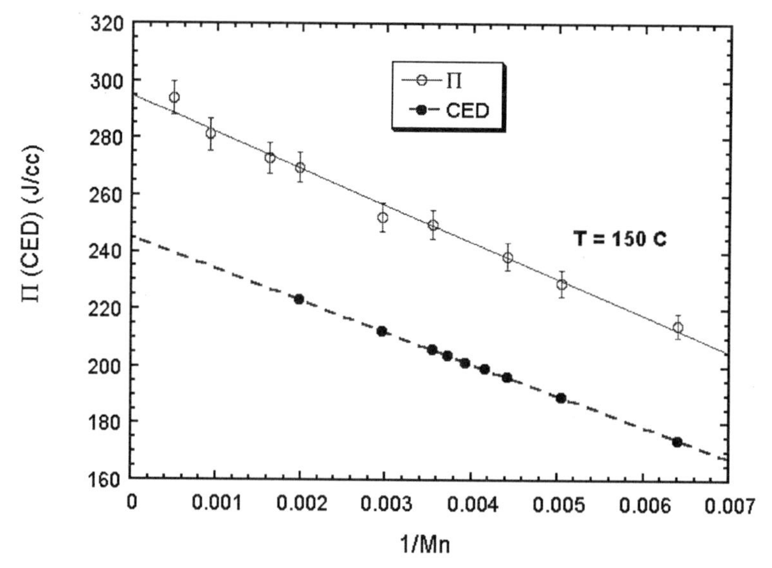

Figure 2: A plot of the internal pressure and the CED as a function of the inverse molecular weight.

polymer set, we can use it to compute the functions n_i for each of the polymer families. In order to accomplish this task we must find a way to determine the true CED for the molecules in one of these data sets. It is to this task we next turn our attention.

The CED for a Polyethylene Oligomeric Series.

Of all the polymers studied, polyethylene (PE) and its oligomers have been most extensively studied.[10,11] The CED for low molecular weight oligomers exists for molecules up to 20 carbon atom chains (C20) over a wide temperature range. The internal pressure can also be measured for these and higher molecular weight liquids.[11] How can we use this data to determine the CED's for the higher molecular weight polyethylene's? A simple and well known property of moleculer weight series of this type is the fact that most if not all bulk thermodynamic properties exhibit a linear dependence on the inverse molecular weight, Mn, for higher molecular weights.[14] The range of molecular weight over which this linear dependence is observed depends on the temperature, pressure and the nature of the thermodynamic property considered.[11] Figure 2 shows a plot of the CED and the internal pressure for the PE series at 150 C. At this temperature a clear linear dependence is observed and one can use these lines to calculate the CED for the higher molecular weight PE's and to compute the value of n_i as a function of molecular weight at this temperature. By repeating this at different temperatures we can construct the function n_i (T,Mw) for polyethylene. Figure 3 shows this function for a number of different molecular weights where the calculation either did not involve extrapolation or where that extrapolation was minimal. Because of the information used in this evaluation, in particular the compressibility, a high degree of precision is not possible. However the behavior of the funcion n is consistant with previous observations[7] and gives us a path foward to the next step in the calculation. Figure 3 shows an asymptotic approach to a constant value of n close to 1.2 as a function of molecular weight and a weak to non-existent temperature dependence of n for these oligomer and polymer polyethylenes. Only values of n below the boiling points of the liquids were included in the figure. Above the boiling point the value of n rises rapidly as one approaches the critical point. Since we are only interested in the liquids well below their boiling points, we have excluded these data points.

The Master Curve for the Surface Tension using the CED.

It is now a simple matter to reconstruct a master curve using the following thermodynamic information set (γ, v_{sp}, α, CED,T, P). Once again we can use the equation of state parameters to fit this data. To insure that the fitting parameters reflect the information set exactly, we do a piece-wise fit to the data as a function of the temperature. We found that most equations of state could not provide a global fit to the data sets with just one set of parameters.[10] Figure 4 shows the results of this scaling using the new information set. What is clear from figure 4 is that we appear to have lost the strong scaling behavior we observed in figure 1, i.e. the data shows an increased deviation from a master curve. As we stated above, a complete collapse of the data would be expected if the only contribution to the surface tension came from the local cohesive forces between the molecules. Hong and Noolandi[15,16] showed that for long chain molecules we expect an additional entropic contribution to the surface tension. Therefore, for a series of molecules with increasing molecular weight, we expect to see a transition from a universal curve for small molecules to a curve which describes long chain molecules. This is what is described in figure 4. Thus there are two universal curves with

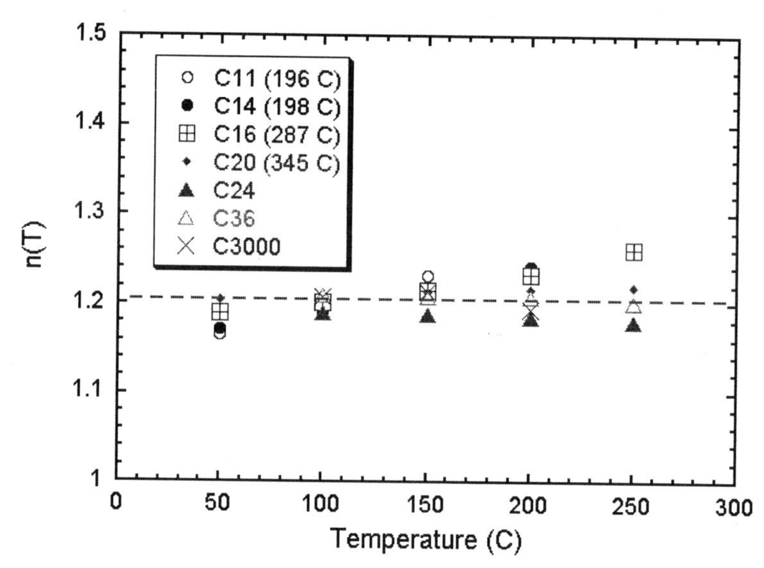

Figure 3: A plot of the function n(T) for polyethylene oligomers as a function of the temperature. The molecular weight of the liquids is indicated in the legend by the number of carbon atoms in the molecule.

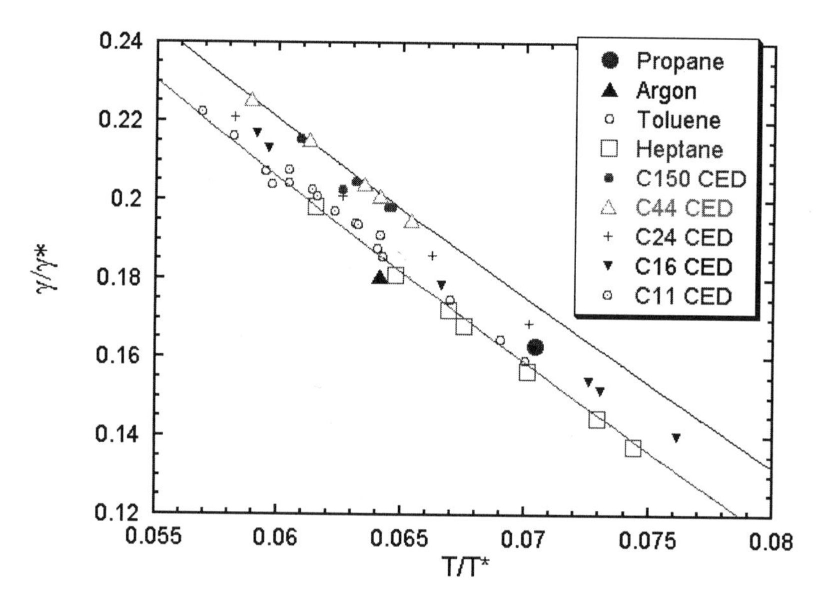

Figure 4: A plot of the scaled surface tension, γ/γ^*, as a function of the scaled temperature T/T^* for polyethylene oligomers and some low molecular weight liquids. γ^* and T^* are defined in figure 1.

the difference between them reflecting the entropic contribution to the surface tension due to the loss of configurations available to chain-like molecules at the liquid/vapor interface. To emphasize this fact, we calculated the scaled surface tensions of a number of small molecule liquids below their boiling points and this data is included in figure 4 where we have fitted curves to the high and low molecular weight data. As expected the PE oligomers lie between the curves unless the molecular weight is high enough so that the entropic contribution is complete. The upper curve can now be used as our reference curve for polymer surface tension data.

Using the polymer master curve we can now use it to compute the functions n_i for different polymers for which we have PVT and surface tension data. The functions n_i define the mapping between the polymer master curves shown in figures 2, which where constructed using the information set (γ, v_{sp}, α, β,T, P), and the master curve shown in figure 4 which was constructed using the information set (γ, v_{sp}, α, CED,T, P). The assumption that there is an exact scaleing of the surface tension with the information set (γ, v_{sp}, α, CED,T, P) is not proved and can only be verified by repeating the analysis described above for PE using other polymer oligomers. The better we describe this master curve the more accurate the predictions will be for the functions n_i and hence the better our predictions of the true CED's of the various polymer liquids will be.

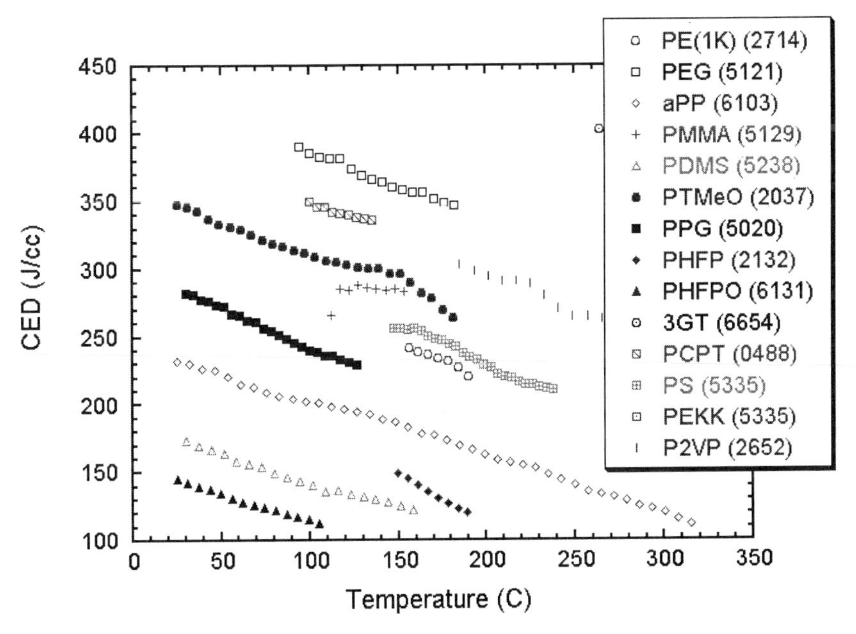

Figure 5: A plot of the CED for a number of different polymers as a function of the temperature. The polymer types are indicated in the legend where the abreviations have the following meaning. PE, polyethylene; PEG, poly(ethylene oxide); aPP, atatic polypropylene; PMMA, poly(methyl methacrylate); PDMS, poly(dimethyl siloxane); PTMeO, poly(tetra-methylene oxide); PPG, poly(propylene oxide); PHFP, poly(hexafluoro propylene); PHFPO, poly(hexafluoro propylene oxide); 3GT, poly(propylene terephthalate); PCPT, polycaprolactone; PS, polystyrene; PEKK, poly(ether ketone ketone); P2VP, poly(2-vinylpyrrolidone).

We computed the values of n_i for a number of polymers with different chemical structures. As observed by Hildebrand and others, we observe that fluoropolymers with weak dispersive interactions have large values of n, and more polar molecules with stronger interactions have values close to 1. Polymers with weaker interactions, fluoropolymers for example, increase with temperature, and polymers with strong polar interaction exhibit weak temperature dependence. Clearly these data pose many interesting questions concerning the relationship between this function n and the chemical structure of the molecules. This is a subject of ongoing research. Figure 5 shows the CED's for these same polymers computed using the above analysis. While the accuracy of the predicted values is still in question, we feel that this approach offers the best route to realistic values of the CED for polymer systems. The values of the CED can be used for phase studies and other calculations where one needs the CED. We can construct thermodynamic tables for enthalpies and entropies for the polymer liquid state relative to an ideal gas state. In cases where we compute enthalpies of mixing, we can hope that we are computing the main contribution to the enthalpy of mixing correctly, and we can hope to dispense with some of the adjustable parameters used in the application of these theories.

References

(1) D.W. van Krevelen "Properties of Polymers, Their Estimation and Correlation with Chemical Structure," Elsevier, Amsterdam, (1976).

(2) J. H. Hildebrand, R. L. Scott; " *The Solubility of Nonelectrolytes*," 3rd Edition, Reinhold Publishing Corporation, (1950).

(3) J. H. Hildebrand, J. M. Prausnitz, and R. L. Scott; " *Regular and Related Solutions*," Van Nostrand Reinhold Company, New York (1970).

(4) G. T. Dee, and B. B. Sauer, *Advances in Physics,* Vol. 47, No. 2, 161-205, (1998).

(5) P. Zoller. In H. Mark, N. Bikales, C. Overberger, and G. Menges, eds., "*Encyclopedia of Polymer Science and Engineering*," Vol. 5, Wiley, New York, 2nd Edition., p. 69, (1986).

(6) B. B. Sauer, N.V. DiPaolo, *J. Colloid Interface Sci.*,144:527 (1991).

(7) G. I. Allen, G. Gee, and G. J. Wilson, *Polymer*, **1**(4), 456 (1960).

(8) P. J. Flory, R. A. Orwoll, and A. J. Vrij, *Am. Chem. Soc.*, 86:3507 (1964).

(9) I.C. Sanchez and R.H. Lacombe, *J. Phys. Chem.*, 80:2352 (1976).

(10)) G. T. Dee and B. B. Sauer, *J. Colloid Interface Sci.* , 152:85 (1992).

(11) G.T. Dee and D.J. Walsh, *Macromolecules*, 21: 811 (1988).

(12) D. Patterson and A.K. Rastogi, *J. Phys. Chem.*, 74:1067 (1970).

(13) G. T. Dee and B. B. Sauer, *POLYMER*, Vol. 36, No.8, 1673, (1995).

(14) T.G. Fox and P.J. Flory, *J. Polym. Sci.*, 14:315 (1954).

(15) K. M. Hong and J. Noolandi, *Macromolecules*, 14:1223 (1981).

(16) K. M. Hong and J. Noolandi, *Macromolecules,* 14:1229 (1981).

THERMAL-DIFFUSION DRIVEN CONCENTRATION FLUCTUATIONS IN A POLYMER SOLUTION

J.V. Sengers[1,2], R.W. Gammon[1], and J.M. Ortiz de Zárate[3]

[1]Institute for Physical Science and Technology
University of Maryland, College Park, MD 20742, USA
[2]Department of Chemical Engineering
University of Maryland, College Park, MD 20742, USA
[3]Facultad de Ciencias Físicas
Universidad Complutense, 28040 Madrid, Spain

INTRODUCTION

Concentration fluctuations in polymer solutions that are in thermal equilibrium are well understood. The intensity of the polymer concentration fluctuations is proportional to the osmotic compressibility and the fluctuations decay exponentially with a decay rate determined by the mass-diffusion coefficient D. Probing these fluctuations with dynamic light scattering provides a convenient way for measuring this diffusion coefficient D [1].

It turns out that the nature of the concentration fluctuations is qualitatively different when polymer solutions are brought into stationary nonequilibrium states. Fuller *et al.* [2] have discussed the enhancement of the concentration fluctuations in polymer solutions subjected to external hydrodynamic and electric fields. In this paper we consider the enhancement of the concentration fluctuations when a polymer solution is subjected to a stationary temperature gradient in the absence of any (convective) flow, that is, while the polymer solution remains in a macroscopically quiescent state. Experimentally this situation can be realized by arranging for a horizontal layer of the polymer solution heated from above.

As originally predicted by Kirkpatrick *et al.* [3] and confirmed by subsequent light-scattering experiments [4,5], a temperature gradient in a liquid induces enhanced temperature and viscous fluctuations that become very long ranged as reviewed by Dorfman *et al.* [6]. The effect of a temperature gradient on the concentration fluctuations in liquid mixtures was originally considered theoretically by Law and Nieuwoudt [7] and has been investigated by Segrè *et al.* [8], Li *et al.* [9] and subsequently by Vailati and Giglio [10]. Here we consider the effect of a temperature gradient on the concentration fluctuations in polymer solutions. The advantage of polymer solutions is that the concentration fluctuations provide the dominant contribution to the Rayleigh component of the scattered light and can be readily measured experimentally.

THEORY

Expressions for the nonequilibrium fluctuations in liquids can be obtained on the basis of fluctuating hydrodynamics [11,12]. Fluctuating hydrodynamics assumes that the fluctuations in

Computational Studies, Nanotechnology, and Solution Thermodynamics of Polymer Systems
Edited by Dadmun *et al.*, Kluwer Academic/Plenum Publishers, New York, 2000

the local temperature T, the local fluid velocity $\mathbf{u}$, and the local concentration c satisfy the linearized hydrodynamic equations supplemented with random force terms [7,12]. Crucial in its application to fluctuations in nonequilibrium states is the assumption that the correlation functions of the random force terms retain their local equilibrium value. The excellent agreement obtained between theory and experiment for the nonequilibrium temperature and fluctuations in one-component liquids has confirmed the validity of this approach [5,9]. Hence, it is expected that fluctuating hydrodynamics provides also a suitable approach for dealing with nonequilibrium concentration fluctuations, although it should be noted that complete agreement between theory and experiment in the case of liquid mixtures has not yet been obtained [9, 13,14].

Law and Nieuwoudt [7] first derived an expression for the concentration fluctuations in a liquid mixture subjected to a temperature gradient from fluctuating hydrodynamics. The theory was further developed by Segrè and Sengers [15] to include the effects of gravity and by Vailati and Giglio [16] to include time-dependent states. The theory can also be extended to colloidal suspensions [17] and to polymer solutions provided that in the latter case the solutions have concentrations where entanglement effects on the hydrodynamics equations can be neglected [18].

A temperature gradient ∇T induces a concentration gradient ∇w through the Soret effect such that [19,20]

$$\nabla w = -S_T w(1-w)\nabla T, \tag{1}$$

where S_T is the Soret coefficient and where w is the concentration expressed as mass fraction of the polymer. Fluctuating hydrodynamics predicts that the concentration fluctuations will couple with the transverse-velocity fluctuations through the induced concentration gradient. This coupling leads to an enhancement of the concentration fluctuations when the transverse-velocity fluctuations have a component in the direction of the concentration gradient. Hence, the nonequilibrium enhancement reaches a maximum for concentration fluctuations with a wave vector $\mathbf{k}$ perpendicular to the concentration gradient. Since the concentration fluctuations do not propagate but are diffusive, they will be insensitive to the upward or downward direction of the perpendicular concentration gradient; the enhancement of the concentration fluctuations will be proportional to the square of the concentration gradient and, hence, proportional to $(\nabla T)^2$ and S_T^2 in Eq. (1). Moreover, as in the case of nonequilibrium temperature and viscous fluctuations observed in liquids and liquid mixtures [8,9], the enhancement of the concentration fluctuations will be inversely proportional to k^4, where k is the wave number of the concentration fluctuations. The divergence of the nonequilibrium enhancement of the fluctuations at small k means that the fluctuations become long ranged. It is now believed that one will always encounter long-range power-law correlations in stationary nonequilibrium states [6]. In practice there will be a low wave-number cutoff due to gravity or to finite-size effects [10,15,21].

The nonequilibrium concentration fluctuations can be measured experimentally by dynamic light scattering. Fluctuating hydrodynamics predicts that the time-dependent correlations function $C(k,t)$ of the scattered light is given by

$$C(k,t) = C_0[1 + A_c(\nabla T, k)]e^{-Dk^2}, \tag{2}$$

where C_0 is the amplitude of the correlation function of the light scattered when the polymer solution is in thermal equilibrium ($\nabla T = 0$) and where $A_c(\nabla T, k)$ represents the enhancement of the amplitude due to the presence of a temperature gradient. As mentioned earlier, this enhancement is proportional to $(\nabla T)^2$ and inversely proportional to k^4:

$$A_c(\nabla T, k) = A_c^* \frac{(\nabla T)^2}{k^4}. \tag{3}$$

In the case that $\mathbf{k} \perp \nabla T$, the coefficient A_c^* becomes [18]

$$A_c^* = \frac{[w(1-w)S_T]^2}{\nu D}\left(\frac{\partial\mu}{\partial w}\right)_{p,T}. \tag{4}$$

In Eq. (4) ν in the kinematic viscosity of the polymer solution, μ the chemical potential difference between solvent and solute such that the derivative $(\partial\mu/\partial w)_{p,T}$, taken at constant pressure p and constant temperature T, is proportional to the inverse osmotic compressibility of the solution.

EXPERIMENTAL OBSERVATIONS OF NONEQUILIBRIUM CONCENTRATION FLUCTUATIONS IN A POLYMER SOLUTION

Light-scattering experiments in solutions of polystyrene with mass-averaged molecular weight $M_W = 96{,}400$ dissolved in toluene have been performed in our laboratory. The polystyrene solution was contained in a horizontal layer located between an upper and lower plate. The optical arrangement was similar to the one used earlier for measuring nonequilibrium fluctuations in liquid mixtures of toluene and n-hexane [9]. Because of the dependence of the nonequilibrium enhancement of the fluctuations on k^{-4}, the experiments have to be done with small values of the scattering angle. Temperature gradients were applied by raising the temperature of the upper plate and lowering the temperature of the lower plate symmetrically so that the average temperature of the solution remained always at 25°C.

Details concerning the experimental apparatus and the experimental procedure will be presented elsewhere [18] The polystyrene solutions used in the experiment were taken from the same batches as those used by Zhang *et al.* [22] for measuring the mass-diffusion coefficient and the Soret coefficient of these polymer solutions.

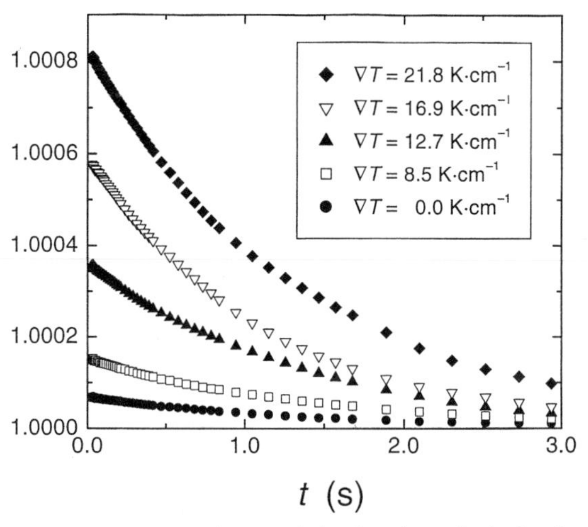

Figure 1. Normalized experimental light-scattering correlation functions, obtained at $k = 1030\ \mathrm{cm}^{-1}$ for a solution of polystyrene ($M_W = 96{,}400$, $w = 2.50\%$) in toluene subjected to various temperature gradients ∇T [18].

Figure 1 shows the experimental time-dependent correlation functions obtained at $k = 1030\ \mathrm{cm}^{-1}$ for various values of the temperature gradient ∇T in a solution of polystyrene in toluene with a polymer concentration of $w = 2.50\%$. The data are relative to the intensity of the stray light that serves as a local oscillator in the heterodyne scattering experiments. The correlation functions obtained for all values of ∇T can be represented by a single exponential

$\propto e^{-Dk^2 t}$ with a diffusion coefficient D independent of ∇T [18]. Hence, the scattering does arise from concentration fluctuations at all values of ∇T. By comparing the experimental correlation functions obtained at finite ∇T with the one obtained at $\nabla T = 0$, we can deduce from the light-scattering measurements an experimental value for the enhancement $A_c(\nabla T, k)$ in Eq. (2).

Experimental data were obtained for solutions with polymer concentrations varying from $w = 0.50\%$ to $w = 4.00\%$ or, equivalently, from $c = 0.00431$ to $c = 0.0348$ g cm^{-3}, to be compared with the estimated overlap concentration $c^* = 0.0274$ g cm^{-3} [22]. Hence, the experimental results correspond to polystyrene-toluene solutions in the dilute and semidilute solution regime.

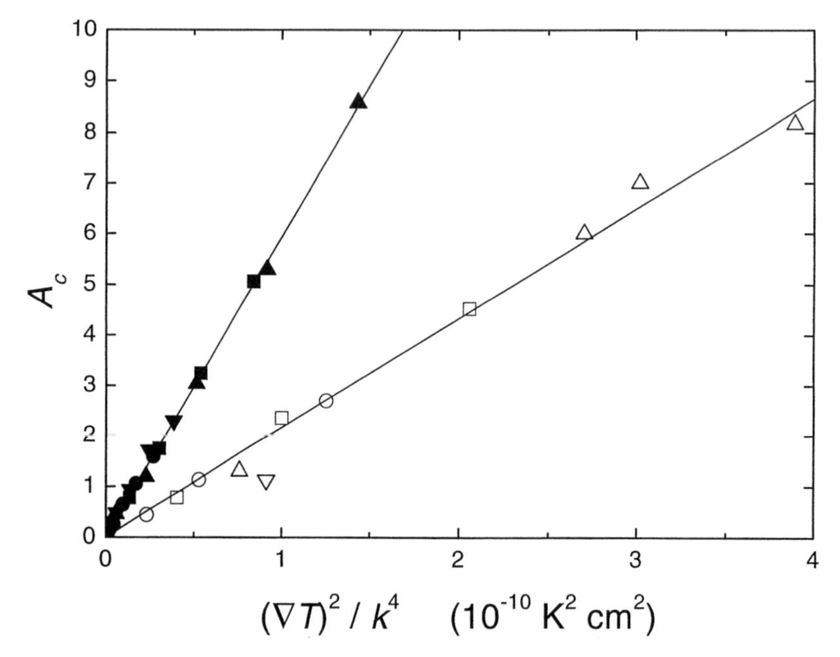

Figure 2. Nonequilibrium enhancement A_c of the concentration fluctuations in a dilute (open symbols: c = 0.00431 g cm^{-3}) and a semidilute (filled symbols: c = 0.00348 g cm^{-3}) polystyrene-toluene solutions as a function of $(\nabla T)^2/k^4$. The lines represent linear fits to the experimental data [23].

In Figure 2 we have plotted the nonequilibrium enhancement A_c of the concentration fluctuations in a dilute and a semidilute polystyrene-toluene solution as a function of $(\nabla T)^2/k^4$ [23]. It is seen that the nonequilibrium enhancement A_c indeed is proportional to $(\nabla T)^2$ and inversely proportional to k^4 in accordance with Eq. (3). Linear fits to these data yield experimental values for the coefficient A_c^* in Eq. (3).

MASS DIFFUSION AND THERMAL DIFFUSION

From Eq. (4) we note that the nonequilibrium enhancement strongly depends on the value of the Soret coefficient. Since the information on the Soret coefficient S_T available in the literature is limited, it was decided to measure S_T of polystyrene-toluene solutions in our laboratory as well [22]. For this purpose an optical beam-deflection method was used. This method was first applied by Giglio and Vendramini [24] to polymer solutions and binary liquid mixtures near a critical mixing point. The method was subsequently improved by Kolodner *et al.* [25] and by Zhang *et al.* [14] for measuring mass diffusion and thermal diffusion in liquid mixtures. In this method one observes the deflection of a laser beam propagating horizontally through the solution upon the imposition of a temperature gradient across the liquid layer. The rate of change of the deflection corresponds to the rate by which a concentration gradient is established and it yields the mass diffusion coefficient D, also referred to as collective diffusion coefficient D_c, of a polymer solution. The final deflection measures the final concentration gradient ∇w resulting from the imposed temperature gradient ∇T and, hence, yields the Soret coefficient in accordance with Eq. (1). The experimental measurements are in good agreement with the measurements obtained by Köhler *et al.* [26] from a forced Rayleigh-scattering method. For details the reader is referred to the paper of Zhang *et al.* [22].

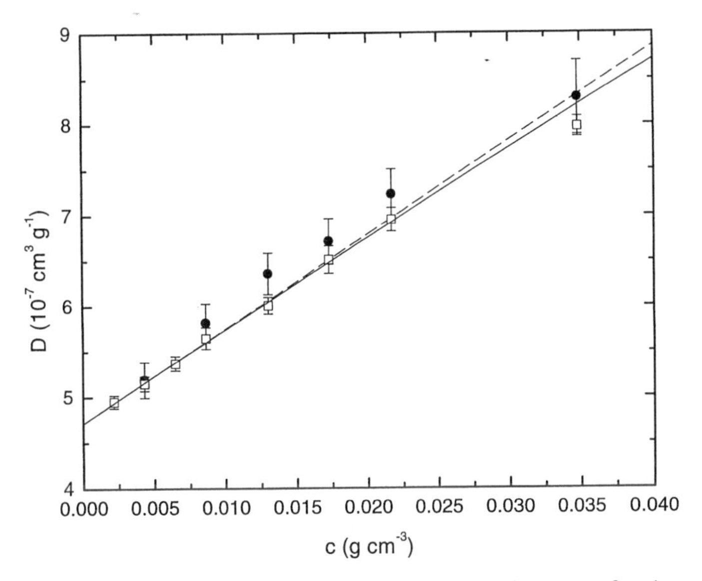

Figure 3. Mass diffusion coefficient D of the polystyrene-toluene solutions as a function of the polymer concentration. The open symbols indicate data obtained with an optical beam-bending technique [22] and the filled symbols data obtained from light scattering. The dashed line represents Eq. (5) and the solid curve represents Eq. (7).

A comparison of the values of the diffusion coefficients deduced from the light-scattering measurements and those measured with the optical beam-bending method is shown in Fig. 3. Zhang *et al.* [22] found that the concentration dependence of the diffusion coefficient is well represented by a commonly used truncated virial expansion [27]:

$$D = D_0(1 + k_D c) \tag{5}$$

with $D_0 = 4.71 \times 10^{-7}$ cm^2 s^{-1} and $k_D = 22$ cm^3 g^{-1}. This linear concentration dependence corresponds to the dashed line in Fig. 3. For concentrations $c/c^* \geq O(1)$ in the semidilute regime, the diffusion coefficient is expected to vary as

$$D \propto (c/c^*)^{\nu/(3\nu-1)} \tag{6}$$

independent of the molecular mass of the polymer [28]. In Eq. (6) $\nu \cong 10/17 \cong 0.588$ is the exponent that characterizes the scaling of the radius of gyration with the mass of the polymer [29]. To combine Eqs. (5) and (6) Nyström and Roots [30] have proposed a scaling approximation of the form

$$\frac{D}{D_0} = \frac{1 + A_D X_D (1 + X_D)^B}{(1 + X_D)^A} \tag{7}$$

with $X_D = r_D(k_D c)$ as a scaling variable and where $A_D = A + r_D^{-1}$. In Eq. (7) $A = (1 - \nu)/(3\nu - 1) = 0.539$ and $B = (2 - 3\nu)/(3\nu - 1) = 0.309$, while r_D is an adjustable constant. The solid curve in Fig. 3 represents Eq. (7) with $r_D = 0.48$ as determined by us from the measurements of Zhang *et al.* [22] for the polystyrene solution with $M_W = 96,400$. In the concentration range of the light-scattering measurements any deviations from a linear relation are negligibly small.

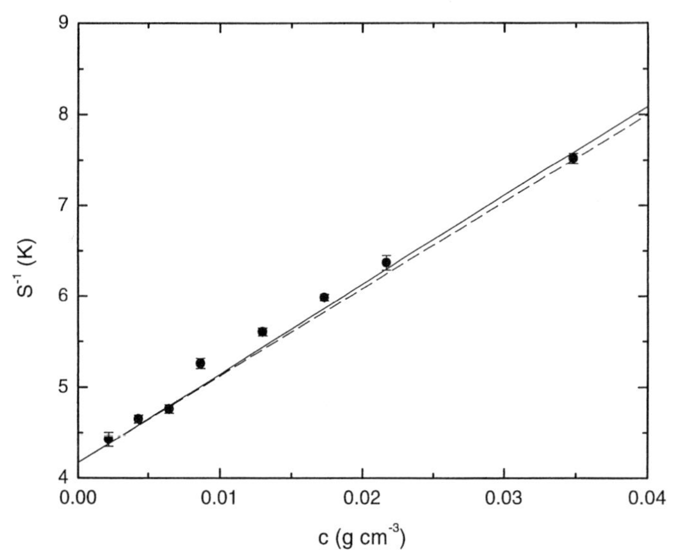

Figure 4. The inverse Soret coefficient S_T^{-1} of the polystyrene-toluene solutions as a function of the polymer concentration. The symbols indicate data obtained with an optical beam-bending technique [22]. The dashed line represents Eq. (9) and the solid curve represents Eq. (10).

The values of the Soret coefficient S_T obtained by Zhang *et al.* [22] for the same polymer solutions in the concentration range corresponding to the light-scattering measurements are shown in Fig. 4. The thermal-diffusion coefficient D_{th} is defined as the product of D and S_T [19,20]

$$D_{\text{th}} = D S_T. \tag{8}$$

The values of D_{th} deduced from the experimental data for D and S_T are shown in Fig. 5. The thermal-diffusion coefficient D_{th} turns out to be rather insensitive to the concentration c, the variation being no more than 5%. Hence, at low concentrations we expect S_T^{-1} to vary as

$$S_T^{-1} = S_{T,0}^{-1}(1 + k_S c) \tag{9}$$

with $S_{T,0} = 0.240\ \text{K}^{-1}$ and $k_S = 24\ \text{cm}^3\ \text{g}^{-1}$ found by Zhang *et al.* [22]. This linear concentration dependence is represented by the dashed line in Fig. 4. To extend the representation of S_T^{-1} beyond the linear regime we assume in analogy to Eq. (7)

$$\frac{S_{T,0}}{S_T} = \frac{1 + A_S X_S (1 + X_S)^B}{(1 + X_S)^A} \tag{10}$$

with $X_S = r_S(k_S c)$ and $A_S = A + r_S^{-1}$, where r_S^{-1} is again an adjustable constant. The solid curve in Fig. 4 represents Eq. (10) with $r_S = 1.16$ as determined by Li *et al.* [18]. The curve in Fig. 5 represents the thermal-diffusion coefficient calculated as the product of D and S_T as given by Eqs. (7) and (10).

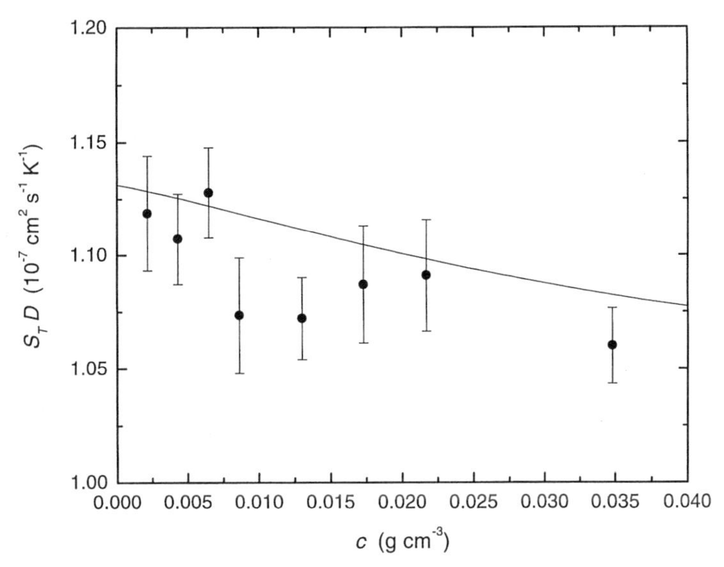

Figure 5. Thermal-diffusion coefficient $D_{\text{th}} = DS_T$ of the polystyrene-toluene solutions as a function of the polymer concentration. The symbols indicate the values deduced from the experimental data obtained for D and S_T with an optical beam-bending technique [22]. The curve represents the values calculated from Eqs. (7) and (10) for D and S_T^{-1}.

COMPARISON WITH THEORY

To check the theoretical prediction for the nonequilibrium enhancement of the concentration fluctuations, as given by Eq. (4), we need in addition to D and S_T also values for the kinematic viscosity ν and for $(\partial\mu/\partial w)_{p,T}$. The latter quantity is related to the concentration derivative of the osmotic pressure [22] which can be deduced from information provided by Noda *et al.* [31,32]. To obtain the kinematic viscosity $\nu = \eta/\rho$ the concentration dependence of the shear viscosity η can be represented by the Huggins relation

$$\eta = \eta_0(1 + [\eta]c + k_{\text{H}}[\eta]^2 c^2) \tag{11}$$

with $\eta_0 = 552.7$ mPa s, $[\eta] = 9.06 \times 10^{-3} M_W^{0.74}$ cm^3 g^{-1} and $k_{\text{H}} = 0.35$, while the density ρ can be calculated from an equation proposed by Scholte [33]. Details are presented in another publication [18].

In Fig. 6 we show the strength A_c^* of the nonequilibrium enhancement of the concentration fluctuations as a function of the concentration c. The circles represent the experimental data and the curve the values calculated from Eq. (4). It is seen that experiment and theory are in satisfactory agreement. We note that the nonequilibrium enhancement of the concentration fluctuations reaches a maximum when the concentration c is approximately equal to the overlap concentration. The nature of the nonequilibrium concentration fluctuations at higher polymer concentration would be an interesting subject for future exploration.

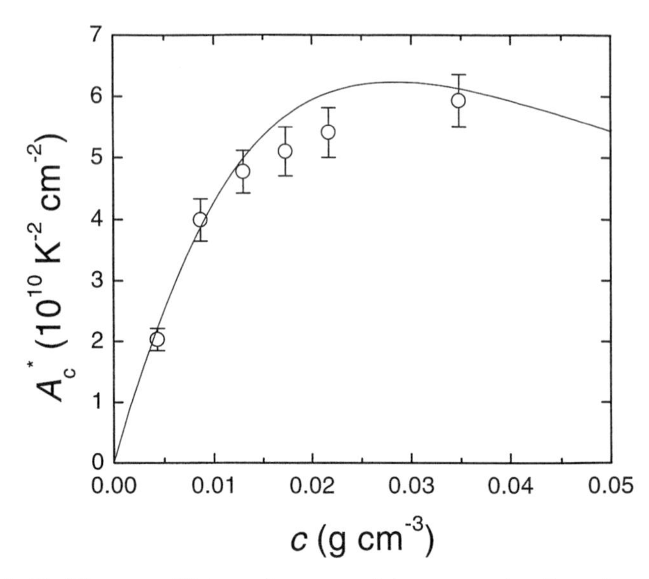

Figure 6. Strength A_c^* of the nonequilibrium enhancement of the concentration fluctuations as a function of the polymer concentration. The symbols indicate experimental data and the curve represents values calculated from Eq. (4).

ACKNOWLEDGEMENTS

The authors are indebted to J.F. Douglas for valuable discussions. The research is supported by the National Science Foundation under Grant No. CHE–9805260.

REFERENCES

1. B.J. Berne and R. Pecora. *Dynamic Light Scattering*, Wiley, New York (1976).
2. G. Fuller, J. van Egmond, D. Wirtz, E. Peuvrel-Disdier, E. Wheeler, and H. Takahashi, in *Flow-Induced Structure in Polymers*, ACS Symposium Series 597, A.I. Nakatani and M.D. Dadmum, eds., American Chemical Society, Washington, DC (1995), p. 22.
3. T.R. Kirkpatrick, E.G.D. Cohen, and J.R. Dorfman, *Phys. Rev. A*, 26:995 (1982).
4. B.M. Law, R.W. Gammon, and J.V. Sengers, *Phys. Rev. Lett.* 60:1554 (1988).
5. P.N. Segrè, R.W. Gammon, J.V. Sengers, and B.M. Law, *Phys. Rev. A* 45:714 (1992).
6. J.R. Dorfman, T.R. Kirkpatrick, and J.V. Sengers, *Annu. Rev. Phys. Chem.* 45:213 (1994).
7. B.M. Law and J.C. Nieuwoudt, *Phys. Rev. A* 40:3880 (1989).
8. P.N. Segrè, R.W. Gammon, and J.V. Sengers, *Phys. Rev. E* 47:1026 (1993).
9. W.B. Li, P.N. Segrè, R.W. Gammon, and J.V. Sengers, *Physica A* 204:399 (1994).
10. A. Vailati and M. Giglio, *Phys. Rev. Lett.* 77:1484 (1996); in *Prog. Colloid. Polym. Sci.* 104:76 (1997).
11. D. Ronis and I. Procaccia, *Phys. Rev. A* 26:1812 (1982)
12. B.M. Law and J.V. Sengers, *J. Stat. Phys.* 57:531 (1989).
13. W.B. Li, P.N. Segrè, J.V. Sengers, and R.W. Gammon, *J. Phys.: Condens. Matter* 6:119 (1996).
14. K.J. Zhang, M.E. Briggs, R.W. Gammon, and J.V. Sengers, *J. Chem. Phys.* 104:6881 (1996).
15. P.N. Segrè and J.V. Sengers, *Physica A* 198:46 (1993).
16. A. Vailati and M. Giglio, *Phys. Rev. E* 58:4361 (1998).
17. R. Schmitz, *Physica A* 206:25 (1995).
18. W.B. Li, K.J. Zhang, J.V. Sengers, R.W. Gammon, and J.M. Ortiz de Zárate, *J. Chem. Phys.* 112:9139 (2000).
19. S.R. de Groot and P. Mazur. *Non-Equilibrium Thermodynamics*, Dover, New York (1984).
20. R. Haase. *Thermodynamics of Irreversible Processes*, Dover, New York (1969).
21. P.N. Segrè, R. Schmitz, and J.V. Sengers, *Physica A* 195:31 (1993).
22. K.J. Zhang, M.E. Briggs, R.W. Gammon, J.V. Sengers, and J.F. Douglas, *J. Chem. Phys.* 111:2270 (1999).
23. W.B. Li, K.J. Zhang, J.V. Sengers, R.W. Gammon, and J.M. Ortiz de Zárate, *Phys. Rev. Lett.* 81:5580 (1998).

SMALL ANGLE NEUTRON SCATTERING FROM POLYMERS IN SUPERCRITICAL CARBON DIOXIDE

George D. Wignall

Solid State Division
Oak Ridge National Laboratory*
Oak Ridge, TN 37831-6393

INTRODUCTION:

Carbon dioxide (CO_2) is nontoxic, nonflammable, naturally occurring, recyclable and inexpensive. Above its critical pressure (P_c = 73.8 bar) and temperature (T_c = 31 ^{O}C), CO_2 has properties intermediate between a gas and a liquid, i.c. high density and low viscosity. Because of these attributes, it has been proposed as an environmentally responsible replacement for the organic and aqueous media used in many solvent-intensive industrial applications. Despite these intrinsic advantages, CO_2 has a significant drawback in that only two classes of polymers (amorphous fluoropolymers and silicones) have been shown to exhibit appreciable solubility at readily accessible temperatures and pressures (T ~ 100 ^{O}C, P ~ 450 bar) [1,2]. Thus, many polymers (e.g. longchain hydrocarbons, waxes, heavy greases, etc.) do not dissolve in CO_2 and this necessitates the use of emulsifying agents to solubilize the "CO_2-phobic" material. Such surfactants are generally amphiphilic (i.e. the different components of the molecule have different solubilities), and it is well known that hydrophobic oil may be solubilized in water by coating the oil droplets with the hydrophilic (water-soluble) component of a detergent. Thus, molecularly engineered diblock copolymer surfactants, consisting of "CO_2 -phobic" and "CO_2 -philic" blocks, have recently been developed [3] for a wide range of applications in liquid and supercritical CO_2. These include the stabilization of polymer colloids during dispersion polymerizations [4], the formation of micelles [5,6], which can solubilize CO_2-insoluble substances [3], and liquid-liquid extractions via the transfer of water-soluble substances from water into a surfactant-rich liquid CO_2 phase. Small angle neutron and x-ray scattering (SANS and SAXS) methods allow the elucidation of the size and shape of both individual polymer chains and supramolecular structures [8,9] in the resolution range, 5-2000Å. Over the past two decades,

* Managed by Lockheed Martin Energy Research Corporation under contract DE-AC05-96OR-22464 for the U. S. Department of Energy.

SANS has emerged as the most powerful technique for studying polymer phase behavior [9] and the self assembly of amphiphiles in aqueous media [8]. In addition, neutron scattering is particularly suited to study the structure of matter under pressure, including supercritical fluids [10], due to the well known high transmission of many of the materials used in the construction of pressure vessels. One other fact that has been used to advantage in the present studies is that fluorinated materials have greater SAXS contrast with the organic core and the solvent. Thus, SAXS and SANS are complementary techniques that highlight different components of the structure, and experiments to apply these techniques to study polymers in SC-CO_2 have recently been undertaken [2,3,5,6,11]. In this publication we review structural information derived from SANS and SAXS studies at Oak Ridge.

EXPERIMENTAL

The SANS data were collected on the W. C. Koehler 30m SANS facility [12] at the Oak Ridge National Laboratory (ORNL) via a 64 x 64 cm^2 area detector with cell size ~ 1 cm^2 and a neutron wavelength, λ=4.75 Å ($\Delta\lambda/\lambda$=0.06). The detector was placed at various sample-detector distances and the data were corrected for instrumental backgrounds and detector efficiency on a cell-by-cell basis, prior to radial averaging, to give a Q-range of $0.005<Q=4\pi\lambda^{-1}\sin\theta<0.05$ Å^{-1}, where 2θ is the scattering angle. The net intensities were converted [13] to an absolute (± 3%) differential cross section per unit sample volume [$d\Sigma(Q)/d\Omega$ in units of cm^{-1}]. The experiments were conducted in the same cell that has been used extensively for polymer synthesis and previous SANS experiments [2,3,5] and the beam passed through two 1 cm sapphire windows, with virtually no attenuation or parasitic scattering. The cross section of the cell filled only with CO_2 amounted to a virtually flat background [5] due to critical scattering (~ 0.04 cm^{-1}), which formed only a minor correction to the scattering from the solutions. SAXS experiments were performed on the ORNL 10m SAXS instrument [14,15], with using $Cu_{K\lambda}$ radiation (λ = 1.54Å) and a 20x20cm^2 area detector with cell (element) size ~ 3mm. Corrections for instrumental backgrounds and detector efficiency have been described previously [7,14,15] and the net intensities were radially averaged in the Q-range $0.009 < Q < 0.055$ Å^{-1} before conversion to an absolute differential cross section [16]. The SAXS cell was similar to that described earlier [11], with single crystal diamond windows (1mm total thickness) and path length ≅ 0.5 mm.

RESULTS AND DISCUSSION

CO_2-Soluble Polymers

For a homogenous polymer solution the methodology to extract the R_g, the radius of gyration (i.e. the r.m.s. distance of scattering elements from the center of gravity) and the second virial coefficient (A_2), which indicates whether a polymer chain swells or contracts in the presence of a solvent is well established [2,9,17]. These parameters are related to the cross section via

$$K\,c\left[\frac{d\Sigma}{d\Omega}(0)\right]^{-1} - \frac{1}{M_W} + 2A_2\,c \qquad (1)$$

where A_2 is the second virial coefficient, M_w is the (weight-averaged) molecular weight, c is the concentration (g cm^{-3}). A_0 is Avogadro's number, and $K = [\Delta(SLD)]^2/\rho_p^2 A_o$ is the contrast factor. Experiments were performed on poly(1,1-dihydroperfluorooctylacrylate) [PFOA] in solution in CO_2. ρ_p is the polymer density and we have assumed that the volume from which CO_2 is excluded by a PFOA chain in the supercritical fluid is the same as the molecular volume in the solid state. [Δ(SLD)] is the neutron scattering length density (SLD) difference between PFOA and CO_2. This gives $K = 7.5 \times 10^{-5}$ mol cm^2 g^{-2} at T = 65 ^{O}C and P = 340 bar. SANS has previously been used [2] to measure A_2 and M_w in the ranges $0.6 < 10^4 A_2 < 0.25$ and $114 < 10^{-3}$ Mw < 1000. Small angle scattering is the most direct method currently available to provide such information in supercritical fluids.

The molecular structure of PFOA is indicated in fig. (1) and polymer chain dimensions were derived using both Zimm plots [fig. (1a)] and also by fitting the data to a Debye random coil model [9], with good agreement between the two approaches. As the pressure is reduced, the solubility decreases and the PFOA falls out of solution at the critical, "neutron cloud point" (T = 65 ^{O}C, P ~ 300 bar), as indicated [fig. (1b)] by a zero intercept [$d\Sigma/d\Omega(0) \Rightarrow \infty$]. The values of R_g are a function of molecular weight (M_w) and may be summarized as $R_g = (0.10 \pm 0.02)M_w^{0.5}$. The second virial coefficient decreases with molecular weight, as generally observed for polymer solutions, where $A_2(M_w)$ is empirically described by $A_2 \sim M_w^{-\delta}$, with $\delta \sim 0.3$ in various systems [18]. From the two SANS data

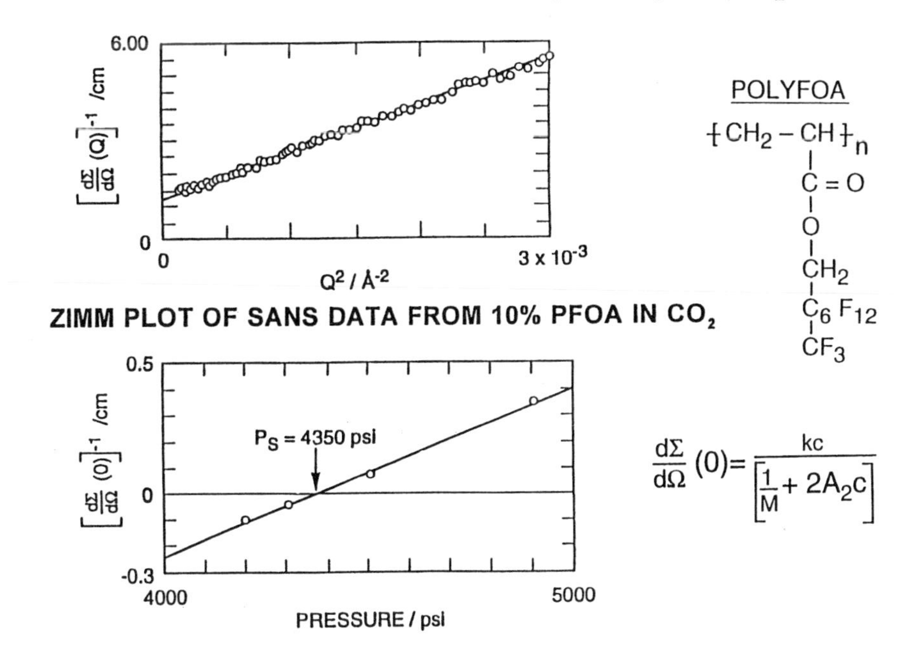

Figure 1. (a). Zimm plot of SANS data for 10% (w/v) high molecular weight POLYFOA in supercritical CO_2; (b). Solubility determination using SANS for 3%(w/v) PFOA in CO_2 at 65 ^{O}C

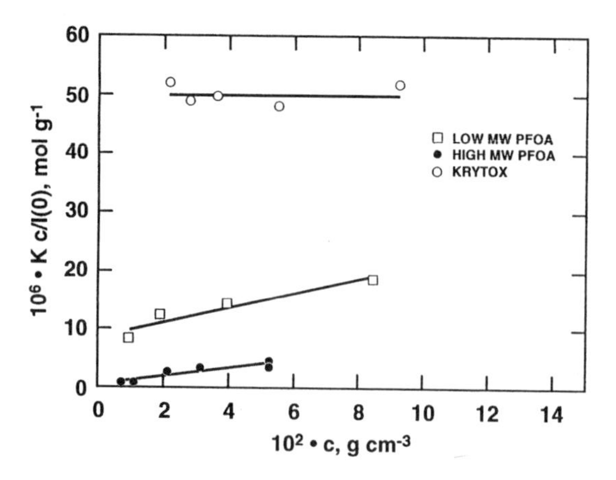

Figure 2. PFOA and HFPPO (Krytox™) in CO_2 at 5000 psi and 65 ^{0}C

points, the exponent is $\delta = 0.4 \pm 0.1$ for PFOA in supercritical CO_2. Thus, within the experimental error PFOA in CO_2 in this range of temperature and pressure behaves similarly to polymers in "good" organic solvents (e.g. polymethyl methacrylate in acetone [18]).

Block Copolymer Micelles

Fig. 2 shows a plot of $Kc[d\Sigma/d\Omega(0)]^{-1}$ vs c for poly(hexafluoropropylene oxide) (PHFPO or Krytox™). The SANS-M_w is in good agreement with the nominal M_w given by the manufacturer (16k/mol) and A_2 is zero within the experimental error, indicating that the polymer coil adopts the unperturbed chain dimensions. In general, $R_g \sim M_w^{0.5}$ and for c < 0.093 g cm^{-2}, R_g is in the range 26-33Å, giving $R_g/M_w^{0.5} \sim 0.24$ Å(g mol)$^{-0.5}$, compared to 0.45 for polyethylene, 0.35 for polypropylene, 0.27 for polystyrene and 0.1 Å (g mol)$^{-0.5}$ for PFOA. Thus, polyethylene molecules have the largest size for a given M_W, because all the CH_2 groups are in the backbone. As the size of the pendant group increases, the fraction of the chain in the backbone decreases. As the radius of gyration depends predominantly on the length and stiffness of the backbone, the overall size of the coil for a given (total) molecular weight becomes smaller.

The SANS data were modelled as a system in which core-shell micelles interact in a solvent medium and, assuming no orientational correlations, the differential scattering cross section is given by

$$\frac{d\Sigma}{d\Omega}(Q) = N_p \left[<|F(Q)|^2> + |<F(Q)>|^2 \ (S(Q) - 1) \right] + B \qquad (2)$$

where N_p is the number density of particles, S(Q) is the structure function arising from interparticle scattering and B is the coherent background (~ 0.04 cm^{-1}) from CO_2. For the dilute solutions considered here, interparticle interactions may be neglected, and the interparticle structure function is similar to that of a dilute gas ($S(Q) \cong 1$). For a core/shell micelle the intraparticle term in Equation (2) may be obtained by integrating over a distribution of core radii with a normalized frequency of $f(R_1)$ and the structure function of a particle with (inner) core radius R_1 and (outer) shell radius R_2 is given by [3]

$$F(Q,R)=\frac{4\pi}{3}[R_1^3(\rho_1-\rho_2)F_o(QR_1)+R_2^3(\rho_2-\rho_s)F_o(QR_2)]$$

$$F_o(x)=\frac{3}{x^3}(\sin x - x\cos x) \tag{3}$$

ρ_1, ρ_2 are the core/shell SLDs and several particle shapes were used to calculate P(Q). The best fits were given by a spherical core-shell model with a Schultz distribution [3] of particle sizes and the aggregation number (i.e. the number of molecules per micelle, N_{agg}), the shell-SLD (ρ_2) and the breadth parameter in the Schultz distribution (Z) were adjusted to fit the data. R_1 and R_2 were calculated from the fitted parameters.

Styrene has been polymerized in CO_2 using polystyrene-b-PFOA (PS-b-PFOA) surfactants, which give rise to quantitative yields (>90%). Samples are designated by the number averaged molecular weights of the blocks (in k Daltons) and figure 3 illustrates the phenomenon of micelle formation for the 3.7k/16.6k PS-b-PFOA copolymer [(4% w/v, 65 ^{O}C, 343 bar (5Kpsi)], caused by the aggregation of the block copolymers in solution. The cross section is much higher than for single molecules, indicating that the particle has a larger molecular mass than the unimer. The subsidiary maximum at Q = 0.04 A^{-1} is related to the spherical shape and the core-shell morphology of the particle. When the PS-block size was kept constant, and the PFOA-block size was increased by a factor of ~ 4, the aggregation

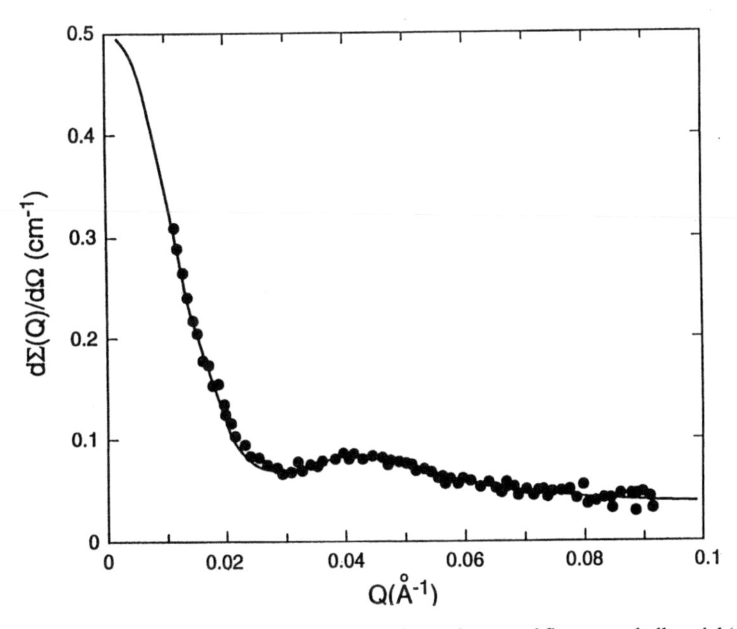

Figure 3. $d\Sigma/d\Omega(Q)$ for 3.7k/16.6k polystyrene-PFOA block copolymer and fit to core-shell model (4% w/v, 65 ^{O}C, 344 bar).

number (N_{agg}) was shown to be relatively independent of the PFOA (corona) block length (MW), as predicted by Halperin and coworkers [19]. Figure 4 shows a comparison of the independently calibrated SAXS and SANS data taken from 3.7k/40k block copolymer solutions at similar experimental conditions and the values of R_1 and N_{agg}, resulting from model fits (Figure 4) are virtually identical. This forms a useful cross check on the methodology, as the contrast factors are quite different for SANS and SAXS. However, SANS has the additional advantage of using isotopic substitution to vary the contrast (see below) and the results [6] indicate that the adjustable parameters are insensitive to changes in concentration, as expected in the "dilute solution" limit. The core radius (R_1) increases with PS-block length and similarly, R_2 appears to increase with the corona block length [6], although the dependence is not strong.

Polystyrene oligomer (M_n = 0.5K g mole^{-1}) was added to the block copolymer solutions to study the solubilization of homopolymers within the micelles. Separate concentration series were run for normal (PSH) and deuterated (PSD) oligomers and the isotopic difference between the oligomers should have a major effect on the scattered

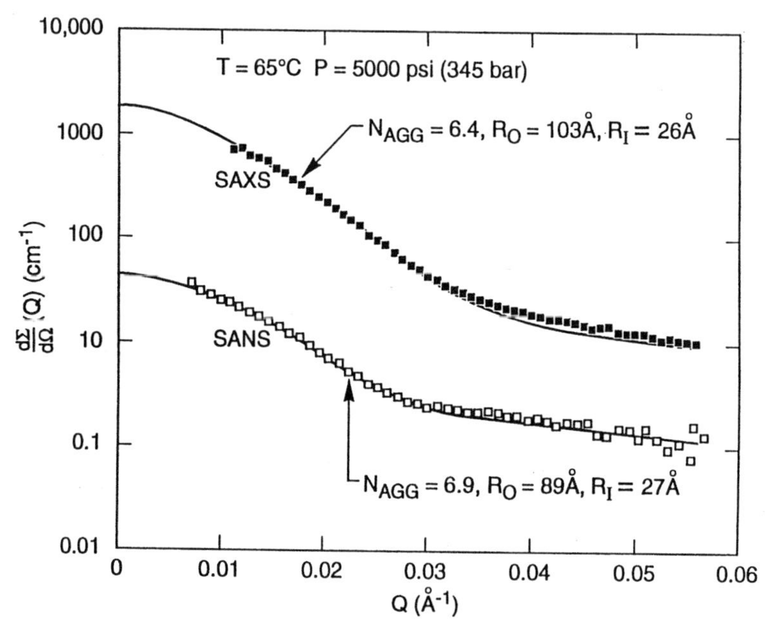

Figure 4. $d\Sigma/d\Omega$ vs. Q for independently calibrated SAXS and SANS data from 3.7k/40k PS-b-PFOA blockcopolymer micelles in CO_2

intensity, as the SLD for PS increases by a factor of ~6 on deuteration. A fourth adjustable parameter was introduced in the model, to account for the fraction of added oligomer solubilized in the core of the micelle (α). The results [6] indicate that virtually all of the added oligomer goes in the core ($\alpha \sim 100\%$), independently of oligomer concentration and isotopic content. Figure 5 shows how the forward SANS cross section increases by over an order of magnitude as the particle size and aggregation number swell to accommodate the polystyrene solubilized in the core. For runs performed at 40 ^{O}C, i.e. higher solvent density, as compared to 65 ^{O}C, the number of molecules per micelle (N_{agg}) decreases to about half of what was observed at 65 ^{O}C. This can be understood [6] as an increase in surface area of block copolymer in contact with the solvent, concomitant with the increase in solubility of the block copolymer at higher density. The data also indicate that an increase in solvent density

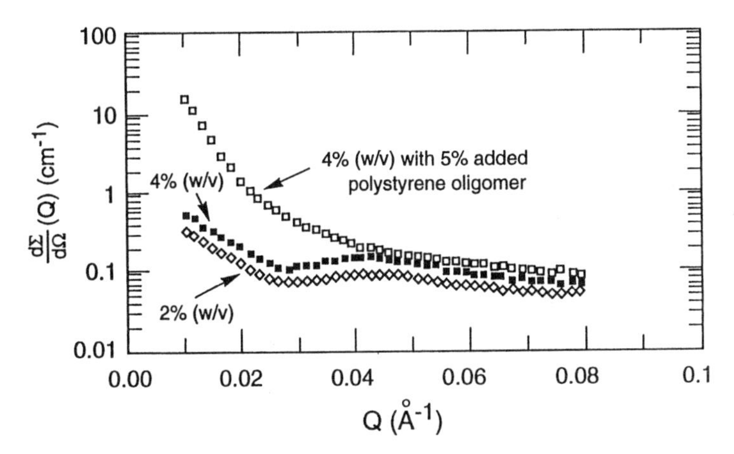

Figure 5. $d\Sigma/d\Omega(Q)$ for 3.7k/16.6k polystyrene-PFOA block copolymer in supercritical CO_2 as a function of concentration (C) of added polystyrene oligomer

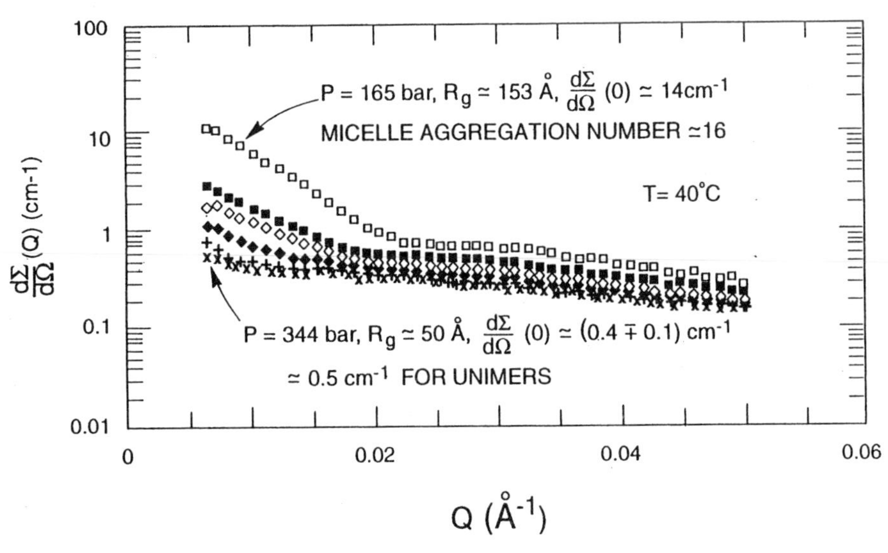

Figure 6. $d\Sigma/d\Omega(Q)$ for 6% (w/v) polyvinyl acetate-b-PFOA block copolymers in CO_2. At high pressures (275 bar), the scattering arises from single molecules; as the pressure is lowered, micelles form below a critical CO_2 pressure.

has the effect of breaking apart a collection of aggregates of relatively low polydispersity (65 ^{O}C), into a collection of smaller aggregates of higher polydispersity. This suggests the existence of a critical micellar density (CMD), which corresponds to the density of the solvent at which the micellar aggregates disappear. Recent experiments [3] on poly(vinyl acetate)-b-PFOA diblock copolymers demonstrate this new phenomenon (Figure 6). The SANS cross section increases by over an order of magnitude as the CO_2 pressure decreases, and thus signals the transition from unimers to micellar aggregates.

The self-assembly of molecules in condensed phases is a ubiquitous phenomenon found in many biological structures, as well as in systems of interest to colloid and surface science. Supercritical CO_2 affects unique perturbations over associative block copolymer systems, as the solvent strength is easily tunable with density and the system may be driven from an aggregated to a dispersed state simply by changing the pressure or temperature to control the solubility. The CMD represents a novel phenomenon, into which neutron scattering provides detailed structural insight. While it is analogous in some respects to other transitions (e.g. changing solvent properties by adding a co-solvent), we believe that the beauty of the CMD is that it is reversible with exquisite control and it is possible to design precisely where this transition will occur. The ability to create and disperse micells, coupled with the fact that CO_2 – phobic materials can be solubilized within the aggregates [3,6,20] makes the CMD an effect that may be of technological importance in the development of environmentally benign processes.

SUMMARY

SANS investigations of individual polymer chains and supramolecular structures have verified the high solubility of amorphous fluoropolymers in CO_2, as well as the formation of micelles by block copolymers when CO_2 is a preferential solvent for one of the blocks. In addition, the results clearly show that the micellar cores are capable of solubilizing CO_2-insoluble material. A transition from unimers to aggregates occurs as a function of pressure, thus demonstrating that solvent strength is easily tunable with changes in solvent density, offering an unprecedented control over the solubility. The CMD constitutes a new concept in colloid and surface chemistry. The self-assembly of molecules in condensed phases is a ubiquitous phenomenon found in many biological structures, as well as in systems of interest to materials, colloid and surface science. These experiments indicate that neutron scattering has the potential to give the same level of microstructural insight into individual polymer molecules (unimers) and colloidal aggregates (micelles) in supercritical CO_2 that it has provided in the condensed state and in aqueous media.

ACKNOWLEDGMENTS

The author wishes to acknowledge his co-workers at Oak Ridge (H. D. Cochran, J. D. Londono, and Y. B. Melnichenko) and at the Universities of North Carolina (D. E. Betts, D. A. Canelas, J. M. DeSimone, J. B. McClain, E. T. Samulski), Palermo (R. Triolo, F. Triolo), Tennessee (R. Dharmapurikar, K. D. Heath, S. Salaniwal), Maine (E. Kiran) and also at the Max Planck Institutut für Polymerforschung (M. Stamm), with whom he has collaborated in many of the studies cited. The research at Oak Ridge was supported by the Divisions of Advanced Energy Projects, Materials and Chemical Sciences, U. S. Department of Energy under contract No. DE-AC05-96OR22464 with Lockheed Martin Energy Research Corporation.

References

1. M. A. McHugh and V. J. Krukonis. *Supercritical Fluid Extraction*, Butterworth – Heinemann, Boston (1986).
2. J. B. McClain, J. D. Londono, T. J. Romack, D. P. Canelas, D. E. Betts, E. T Samulski. J. M. DeSimone, and G. D. Wignall, Solution properties of CO_2-soluble fluoropolymers: a SANS investigation of solvent density on the radius of gyration, *J. Amer. Chem. Soc.* 118: 917 (1996).
3. J. B. McClain, J. D. Londono, D. Chillura-Martino, R. Triolo, D. E. Betts, D. A. Canelas, H. D. Cochran, E. T Samulski. J. M. DeSimone, and G. D. Wignall, Design of nonionic surfactants for supercritical carbon dioxide, *Science* 274: 2049 (1996).
4. J. M. DeSimone, Z. Guan, and C. S. Eisbernd, Synthesis of Fluoropolymers in supercritical carbon dioxide, *Science* 265: 356 (1994).
5. D. Chillura-Martino, R. Triolo, J. B. McClain, J. Combes, D. Betts, D. Canelas, E. T. Samulski, J. M. DeSimone, H.D. Cochran, J. D. Londono, and G. D. Wignall, Neutron scattering characterization of polymerization mechanisms in supercritical CO_2, *J. Molecular Structure* 383: 3 (1996).
6. J. D. Londono, R. Dharmapurikar, H. D. Cochran, G. D. Wignall, J. B. McClain, J. R. Combes, D. E. Betts, D. A. Canelas, J. M. DeSimone, E. T. Samulski, D. Chillura-Martino, and R. Triolo, The morphology of blockcopolymer micelles in supercritical CO_2 by small angle neutron and x-ray scattering, *J. Appl. Cryst.* 30: 690 (1997).
7. A. I. Cooper, J. D. Londono, G. D. Wignall, J. B. McClain, E. T. Samulski, J. S. Lin, A. Dobrynin, M. Rubinstein, J. M. J. Frechet, and J. M. DeSimone, Extraction of a hydrophilic compound from water into liquid CO_2 using dendritic surfactants, *Nature* 389: 368 (1997).
8. J. B. Hayter *Physics of Amphiphiles, Micelles, Vesicles and Microemulsions*, North Holland, Amsterdam (1985).
9. G. D. Wignall, Neutron scattering from polymers, in *Encyclopedia of Polymer Science and Engineering*, Martin Grayson and J. Kroschwitz, eds., John Wiley & Sons, Inc., New York (1987).
10. J. D. Londono, V. M. Shah, G. D. Wignall, H. D. Cochran, and P. R. Bienkowski, Small-angle neutron scattering studies of dilute supercritical neon, *J. Chem. Phys.* 99: 466 (1993).
11. J. L. Fulton, D. M. Pfund, J. B. McClain, T. J. Romack, E. E. Maury, J. R. Combes, E. T. Samulski, J. M. DeSimone, and M. Capel, Aggregation of amphiphilic molecules in supercritical carbon dioxide: a small angle x-ray scattering study, *Langmuir* 11: 4241 (1995).
12. W. C. Koehler, The national facility for small-angle neutron scattering, *Physica (Utrecht)* 137B: 320 (1986).
13. G. D. Wignall, and F. S. Bates, Absolute calibration of small-angle neutron scattering data, *J. Appl. Cryst.* 20: 28 (1987).
14. R. W. Hendricks, The ORNL 10-mater small-angle x-ray scattering camera, *J. Appl. Phys.* 11: 15 (1978).
15. G.D. Wignall, J.S. Lin, and S. Spooner, Reduction of parasitic scattering in small angle x-ray scattering by a three-pinhole collimating system, *J. Appl. Cryst.* 23: 241 (1990).
16. T. P. Russell, J. S. Lin, S. Spooner, and G. D. Wignall, Intercalibration of small-angle x-ray and neutron scattering data, *J. Appl. Cryst.* 21: 629 (1988).
17. *Neutron, X-Ray and Light Scattering*, P. Lindner and T. Zemb, eds., Elsevier, New York (1991).
18. H. Fujita. *Polymer Solutions*, Elsevier, Amsterdam (1990).
19. A. Halperin M. Tirrell, and T. P. Lodge, Tethered chains in polymer microstructures, *Adv. Polym. Sci.* 100: 31 (1992).
20. F. Triolo, A. Triolo, J. D. Londono, G. D. Wignall, J. B. McClain, D. E. Betts, S. Wells, E. T. Samulski and J. M. DeSimone, Critical micelle density for the self-assembly of block copolymer surfactants in supercritical carbon dioxide, *Langmuir* 16(2): 416 (2000).

POLYMER SOLUTIONS AT HIGH PRESSURES: Pressure-Induced Miscibility and Phase Separation in Near-critical and Supercritical Fluids

Erdogan Kiran, Ke Liu and Zeynep Bayraktar

Department of Chemical Engineering
Virginia Polytechnic Institute and State University
Blacksburg, Virginia 24061

ABSTRACT

Pressure-induced miscibility and phase separation constitute the integral steps in a wide range of applications that use supercritical or near-critical fluids as a process or processing medium for polymers. Thermodynamic aspects of miscibility, and the kinetic aspects of the phase separation both play a very important role. Pressure becomes the practical tuning parameter that transforms a fluid or a fluid mixture from behaving like a solvent to one behaving like a non-solvent, thereby inducing miscibility or phase separation. Dynamics of phase separation becomes particularly important if transient (non-equilibrium) structures are to be pinned by proper matching of the process conditions with the onset of transitions in material properties such as the vitrification or crystallization in polymers. This paper provides an overview of factors that influence miscibility and presents the consequences of pressure-induced phase separation in terms of the time scale (kinetics) of new phase formation, the domain growth and structure development in polymer solutions. A time- and angle-resolved light scattering technique combined with controlled pressure quench experiments with different depth of penetration into the region of immiscibility is used to document the kinetics of phase separation and domain growth and identify the crossover from "metastable" to "unstable" region of the phase diagram. This crossover that represents the experimentally accessible spinodal boundary is demonstrated with recent data on pressure-induced phase separation in "poly(dimethylsiloxane) + supercritical carbon dioxide" and "polyethylene + n-pentane" solutions. The paper also describes a unique application of polymer miscibility and phase separation concepts in the preparation of novel polymer-polymer blends that is based on impregnating a host polymer that is swollen in a fluid with a second polymer that is dissolved in the same fluid at high pressures. The dissolved polymer is in-situ precipitated and entrapped in the host polymer matrix by pressure-induced phase separation. The technique opens up new possibilities for

Computational Studies, Nanotechnology, and Solution Thermodynamics of Polymer Systems
Edited by Dadmun *et al.*, Kluwer Academic/Plenum Publishers, New York, 2000

blending of otherwise incompatible polymers, and is demonstrated for blending of polyethylene with poly(dimethylsiloxane) in supercritical carbon dioxide .

INTRODUCTION

Misibility or phase separation in polymer solutions may be induced (*i*) by a change in temperature, or *(ii)* by a change in composition as result of adding a third component such as a solvent or a non-solvent to a system, or as a result of progress of a reaction, or (*iv*) due to a change in pressure.[1-4] These are schematically represented in Figure 1. Even though not shown in the figure, applied fields such as shear may also induce phase separation. Of these, the temperature- and solvent-induced phase separation are the most common methodologies that are used in membrane formation and other processes that lead to porous materials. Reaction induced phase separation is encountered in polymerization, and in particular in systems that undergo crosslinking where a homogeneous system may enter the two-phase region with an increase in degree of polymerization (DP) or conversion. Pressure-induced phase separation (PIPS) is essentially an unavoidable integral step in all processes that employ near-critical or supercritical fluids. This is because at some point in the process the pressure

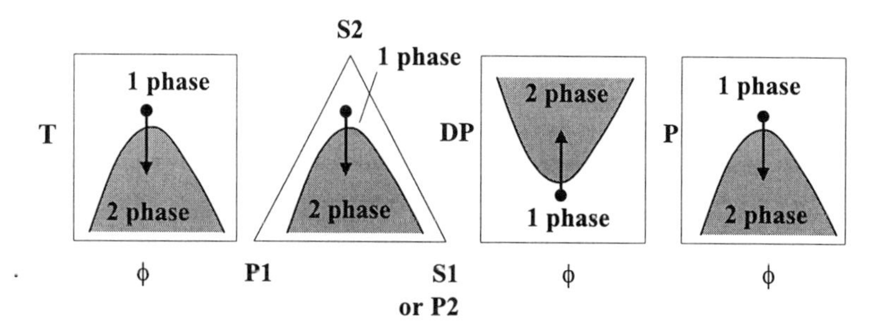

Figure 1. Schematic representation of the influence of various factors on the miscibility and phase separation. The shaded areas represent the two-phase regions. The arrows show the paths for phase separation, while the reverse directions would be the paths for miscibility. From left-to-right: Temperature-induced phase separation; solvent-induced phase separation; reaction (i.e. polymerization) -induced phase separation (the two-phase regions are entered with increasing degree of polymerization (DP)); and pressure-induced phase separation. ϕ = polymer concentration, S2 and S1 solvent and non-solvent; P1 and P2 = polymer 1 and polymer 2.

must be reduced either to recover a product, or the processing fluid. A particularly important aspect of the pressure-induced phase separation that differentiates it from the other techniques is the fact that the pressure changes can be brought about uniformly and very fast throughout the bulk of a solution. This would not be so in other techniques due to, for example, heat and mass transfer limitations that would influence the rate of phase separation in the case of temperature- and solvent-induced phase separations, respectively. It is however important to note that temperature, solvent, reaction, or field-induced phase separation may also be carried out at elevated pressures if desired, and therefore all modes of phase separation methods are of interest when working with near-critical or supercritical fluid systems.

MISCIBILITY

Miscibility of a polymer in a near-critical fluid depends on the temperature, pressure, polymer concentration, molecular weight and molecular weight distribution, the polymer type and the nature of the solvent fluid under consideration. Several comprehensive reviews have recently appeared.[1,5-9] The results of high-pressure miscibility studies are often reported in the form of pressure-temperature demixing curves that are typically based on data generated in a variable-volume view cell. The polymer + fluid mixture corresponding to a target polymer concentration is first brought into the one-phase homogeneous conditions, and then the pressure is slowly decreased. The incipient phase separation condition is visually or optically noted through the windows of the view-cell. Depending upon the polymer + solvent system, the shape of these demixing curves differ. The demixing curves corresponding to liquid-liquid phase separation can be generalized and grouped into the types that are shown in Figure 2. In these diagrams the curves a, b, and c represent three different concentrations and for each concentration the region above each curve is the homogeneous one-phase region. Below each curve is the two-phase region. In Figure 2A, at a given pressure, the system that may be in the two-phase region may become one-phase with increasing temperature (typical of systems showing upper critical solution temperature - UCST), but with further increase in temperature may again enter two-phase region (typical of system showing lower critical solution temperature - LCST). At lower pressures, the two- phase regions may persist at some concentrations resulting in an hour-glass shaped region of immiscibility in the temperature-composition plane. In Figure 2B one-phase regions are entered upon an increase in temperature. The reverse is the case in Figure 2C. The scenario in Figure 2D is not very common and represents the possibility of entering a two phase region first and then becoming one phase at higher temperatures for some pressures which would lead to an island of immiscibility in the temperature-composition plane.

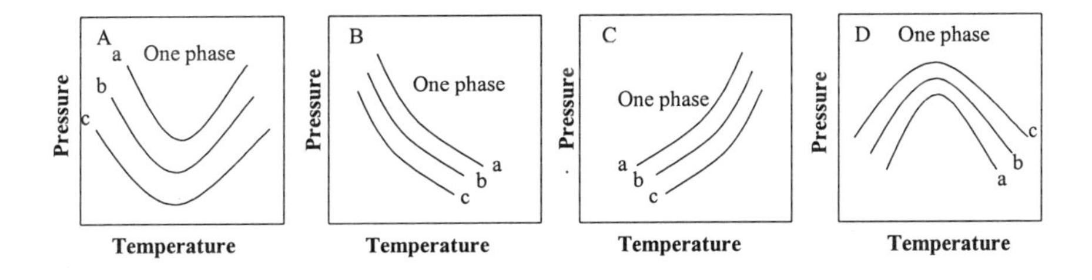

Figure 2. Demixing pressures for different concentrations a, b, c. Below each curve is the two-phase region. Depending upon the pressure, the system may display both UCST and LCST or hour-glass shaped region of immiscibility (A); only UCST (B); only LCST (C), or an island of immiscibility (D) in the temperature-composition plane. (See Figure 3).

Figure 3 shows the temperature-composition diagrams that would result from the constant pressure cuts in Figure 2A, which demonstrates the progressive shift from hour-glass shaped region of immiscibility to observation of both UCST and LCST branches and the eventual widening of the miscible region as pressure is increased. We have recently noted[1,10] the striking similarity between the shape of the P-T demixing curves and the resulting T-x phase diagrams and the temperature dependence of the Flory interaction parameter χ for polymer solutions and the resulting T-x phase diagrams.[11] This analogy between pressure and Flory χ parameter is helpful in understanding the role the pressure plays in influencing the polymer solvent interactions.

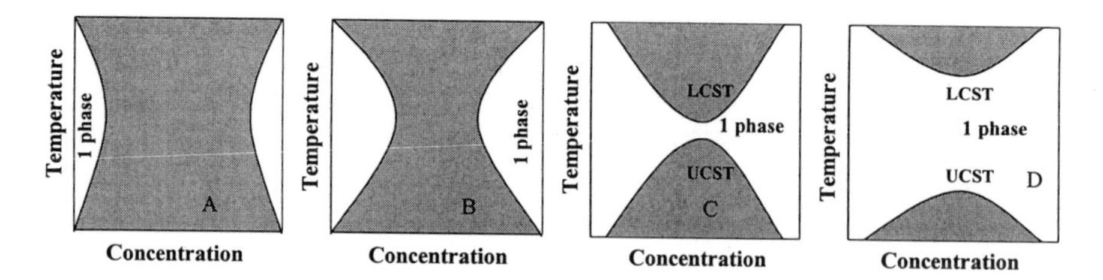

Figure 3. Temperature-composition phase diagrams in polymer solutions that display liquid-liquid demixing pressures of the type shown in Figure 2A. . Shaded areas are the two-phase regions. From left-to-right the system shows progressively improving degree of miscibility as the system moves from one showing an hour glass-shaped region of immiscibility (A and B) to one displaying upper and lower critical solution behavior with widening region of miscibility (C and D) with increased pressure.

A well known example of a polymer - solvent system showing behavior like Figure 2A is the poly(dimethylsiloxane) solutions in supercritical carbon dioxide.[10] Figure 4a shows the demixing pressures for this system for concentrations in the range from 0.06 to 5.15 % by weight. Figure 4b is the corresponding pressure-composition diagram generated at 350 K. These diagrams provide complete information on the minimum pressures needed to achieve complete miscibility for a given polymer concentration at a given temperature. A common feature for many polymer solutions is the relatively flat nature of the P-x diagrams especially for polymers with somewhat broad molecular weight distributions.

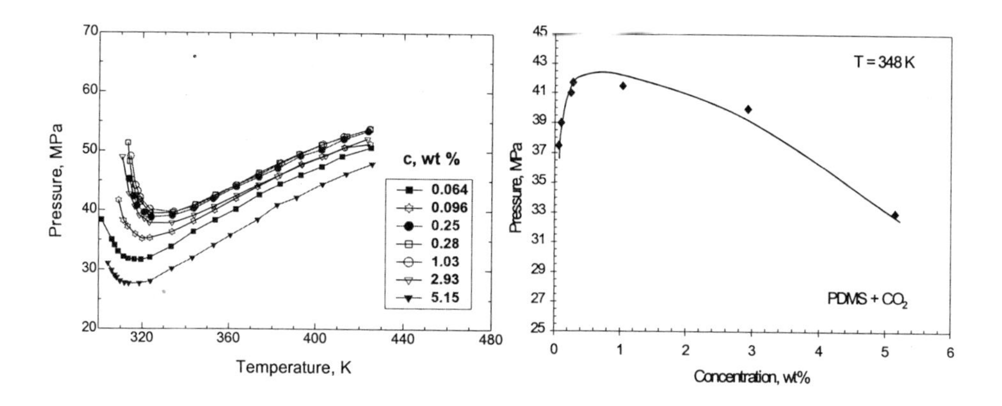

Figure 4. ***A (left).*** Demixing pressures for solutions of poly (dimethylsiloxane) (M_w = 94,300 and M_w/M_n = 3) in supercritical carbon dioxide at different concentrations. ***B (right)*** Pressure-composition phase diagram at 350 K.

Another example of demixing pressures is shown in Figure 5. The figure shows the behavior of polyethylene (M_w = 121,000; PDI = 4.43) in n-pentane and in pentane + carbon dioxide mixtures.[12] In pure pentane (Figure 5A), the system behavior is similar to Figure 2C and shows LCST type behavior. In this figure, the vapor pressure curve for pentane is also included. Most of the miscibility data shown in the figure corresponds to the sub-critical temperatures for pentane. Figure 5B demonstrates how the solvent quality can influence the observed behavior. The figure shows the demixing pressures for a nominal 3 % by wt polyethylene solution in the binary fluid mixtures of pentane and carbon dioxide. The polymer phase behavior in such ternary systems depends strongly on the composition of the solvent fluid. As shown, even though polyethylene shows an LCST type behavior in pure pentane and dissolves at modest pressures, the behavior shifts gradually to a UCST type behavior in pentane + carbon dioxide mixtures with increasing carbon dioxide content, and complete miscibility conditions require very high pressures.[1,12]

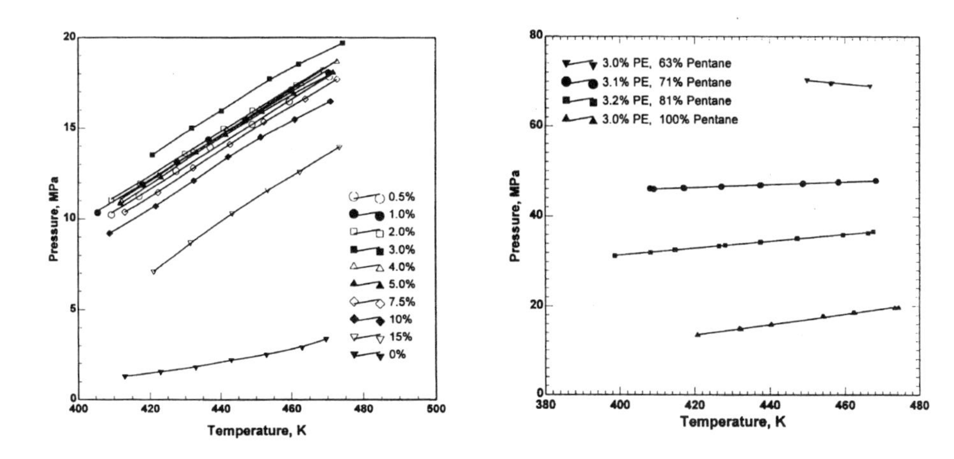

Figure 5. ***A(left)*** Demixing pressures in solutions of polyethylene (M_w= 121,000; PDI= 4.43) in n-pentane at different concentrations. Figure shows the vapor pressure curve for n-pentane also. ***B(right).*** Demixing pressures in 3 % solution of the same polyethylene sample in binary fluid mixtures of pentane and carbon dioxide.

Miscibility in fluid mixtures containing carbon dioxide is of particular importance in applications that aim to reduce the use of traditional organic solvents in polymer processing.[1, 13] For example, use of acetone may be reduced in processing of cellulose acetate by using acetone + carbon dixoide fluid mixtures.[13] A different type of ternary mixtures that is also of particular interest are the mixtures of two different polymers in one common high-pressure fluid. A limiting case of this is the behavior of mixtures of fractions of the same polymer with different molecular weights.[14] Recent studies in our laboratory have shown that mutually incompatible polymers can be mixed and solution blended in supercritical fluids.[1,15] Some subtle but significant effects are however encountered. A particularly interesting case is the miscibility of mixtures of isotactic polypropylene with polyethylene which are mutually incompatible. Both of these polymers are individually soluble in n-pentane at relatively low pressures (i.e., at pressures less than 15 MPa). Their mixtures however, depending upon the actual compositions involved, may require pressures around 50 MPa and higher for complete miscibility in n-pentane.[1,15]

PHASE SEPARATION

Figure 6A is a general representation of the pressure-composition diagrams for systems that display liquid –liquid phase separation. The binodal boundary represents the equilibrium demixing pressures for a monodisperse polymer system. Below the binodal there is another boundary known as the spinodal boundary. The binodal and the spinodal envelopes determine the metastable region (shaded area in between). They merge at the critical polymer concentration (ϕ_c). Starting from a high-pressure in the one-phase region, phase separation may be induced by reducing the pressure. The equilibrium compositions of the two phases that form when pressure is reduced from P_i to P_f are given by the binodals ϕ_{Ib} and ϕ_{IIb}. For the different pressure quench paths I, II and III that end at the same final pressure (P_f), the compositions of the coexisting phases are given by the same binodal compositions, even though the concentrations of the initial one-phase solutions are different. The difference in the initial compositions shows itself in the phase volumes of the coexisting phases rather than the compositions.

Figure 6A is the thermodynamic (equilibrium) picture and does not describe the transient structures that may form along the different pressure reduction paths before final equilibrium conditions are reached. Along the paths I and III, the system enters the metastable regions where the mechanism of phase separation is by nucleation and growth.[1,3-5] In solutions of low polymer concentrations along path I, the polymer-rich phase nucleates in the polymer-lean phase, and grows. In contrast, in the more concentrated polymer solutions along path III, the polymer-lean phase nucleates in the polymer–rich phase and grows. The particles of polymer-rich phase (path I), or the bubbles of polymer-lean phase (path III) unless frozen may eventually collapse, settling into the two coexisting phases with compositions fixed by the binodal values. These are demonstrated in Figure 6B. Along the path II which corresponds to the critical polymer concentration, the system immediately enters the spinodal region where the phase separation is spontaneous and in the initial stage a co-continuous texture of polymer-rich and polymer-lean phases forms. In time, the structure coarsens and may eventually collapse. At concentrations away from the critical polymer concentrations, entering the spinodal domain may not be readily achievable. This depends on the width of the metastable region and the speed with which it can be traversed.

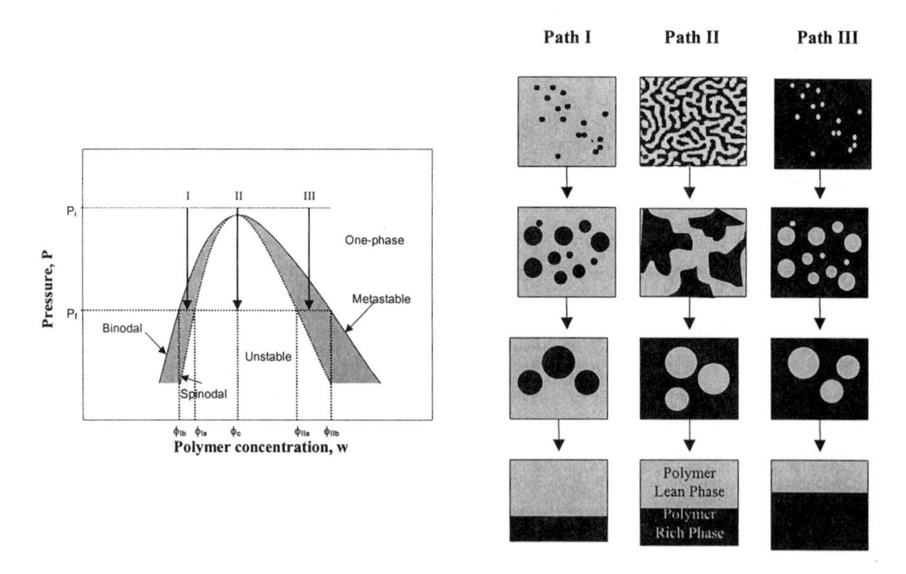

Figure 6. ***A (left)*** Pressure – composition phase diagram. ϕ_{Ib} and ϕ_{IIb} represent the coexistence phase compositions when pressure is reduced from P_i to P_f and the solutions phase separates into a polymer-lean and polymer-rich phase. ***B(right)*** Transient structure formation and evolution when the nucleation and growth (paths I and III) and spinodal decomposition (path II) regimes are entered.

Understanding the time scale of new phase formation and growth and devising methods to quench and pin a system in one of its non-equilibrium transient states are important considerations in producing micro-structured materials for example structures with specific pore size, distribution or morphologies. The system must be quenched in the appropriate time frame if the transient structures are to be locked-in. By documenting the kinetics of phase separation and mapping the experimentally accessible spinodal boundary, one could fine-tune the end-pressures of the quench to induce phase separation by nucleation and growth or spinodal decomposition, and try to capture the transient structures by matching the vitrification or gelation times with the time scale of the domain growth. For example, such an understanding would help develop methodologies for formation of co-continuous morphology development even from polymer solutions that may not be initially at the critical polymer concentration.

KINETICS OF PHASE SEPARATION

We employ a unique experimental system that has been developed in our laboratory to document the kinetics of pressure-induce phase separation and make assessment on the metastable gap through verification of the crossover from the nucleation and growth to spinodal type phase separation mechanism.[1,2,16-19] The technique combines the notion of multiple repetitive pressure-drop (MRPD) with time- and angle resolved light scattering. Figure 7 illustrates the basic methodology and the essential components of the system. At any given polymer concentration, the system pressure is reduced from a point A in the homogenous one-phase region to a point B while the time evolution of the angular variation of the (I_s) is monitored. The solution is then homogenized and brought back to the initial condition A. Now a second pressure quench is imposed, this time to a lower end-pressure such as the point C while again monitoring the time-evolution of the scattered light intensities. The process is repeated by bringing the solution to its homogenous condition A, and then subjecting the solution to deeper quenches to with end-pressures at points D, E or lower. Depending upon the depth of penetration into the region of immicibility, the time-evolution of the angular variation of the scattered light intensities show characteristic differences when the spinodal boundary is crossed. In the nucleation and growth (NG) regime scattered light intensties show a continual decrease with increasing angle θ (or the wave vector q) while in the spinodal regime (SD) the angular variation of the scattered light intensities display a maximum. These type of experiments are repeated for solutions of different concentrations that span concentrations below and above the critical polymer concentrations to permit generation of the complete pressure-composition phase diagrams with experimentally accessible spinodal boundaries.

The light-scattering cell that is shown in Figure 7 is designed to have a short path length with two flat sapphire windows (SW) that are separated by 250 microns. The cell geometry permits monitoring the scattered light intensities over a range of angles. The cell body has a built-in piston that is moved by a pressurizing fluid (PF) with the aid of a pressure generator, a movable air-actuated rod (PR), and a dedicated valve stem (V) that can be used to impose pressure quenches of different depths and rates. The solution is first homogenized and fully circulated through the scattering cell with a circulation loop (not shown in the figure). Then, the cell is isolated and the internal pressure is changed either by the movement of the piston (for small or large pressure changes at a slow rate), or the valve stem (for small and rapid pressure changes), or the air-actuated rod (for large and rapid pressure changes). The actual temperature and the pressure in the solution as well as the transmitted and the scattered light intensities over an angle range from 0 to 13 degrees are monitored in real time during pressure quench. All angles are scanned every 3.2 ms to generate information of the time evolution of the scattered light intensities at all angles.

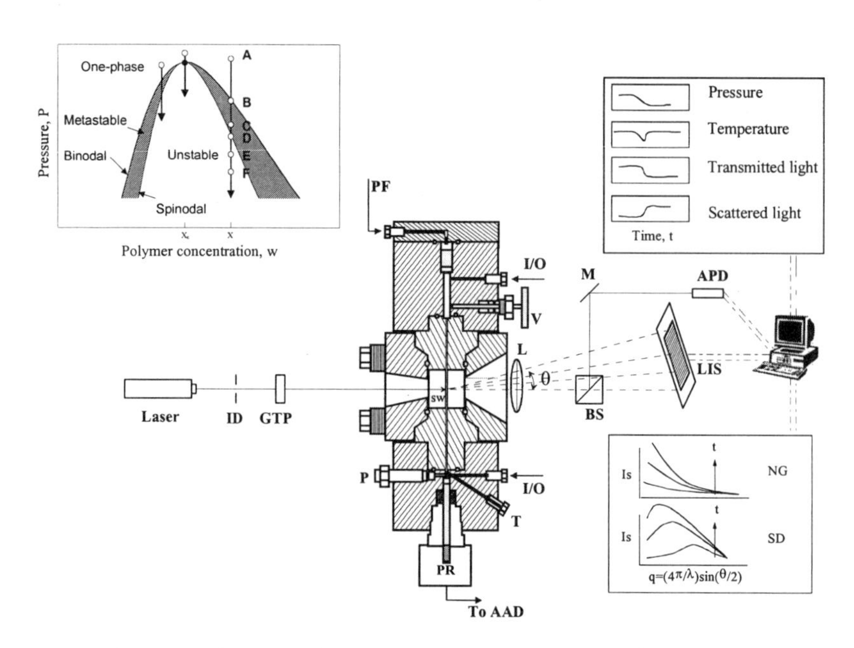

Figure 7. Experimental system and the methodology to study kinetics of phase separation. Controlled pressure quenches are imposed to bring the system from a homogeneous one-phase state into metastable and unstable regions while changes in pressure, temperature, transmitted light and scattered light at a range of angles are monitored in real time. The angular variation of the scattered light intensities during phase separation shows characteristic fingerprint patterns depending upon the mode of phase separation. Nucleation growth (NG) lead to continual decay, but spinodal decomposition (SD) lead to a maximum in the scattered light intensity profiles.

Using this system we have already investigated the kinetics of pressure-induced phase separation in several polymer-solvent systems including "polystyrene + methyl cyclohexane" [17], "poly (dimethylsiloxane) + supercritical carbon dioxide" [18], and "polyethylene + n-pentane".[19] Figure 8 shows the time-evolution of the scattered light intensities for a 5.5 wt % PDMS (M_w= 94,300, M_w/ M_n = 2.99) solution in supercritical carbon dioxide in two different quench experiments. The angular variation is expressed in terms of the wave number which is given by $q = (4\pi / \lambda) \text{Sin}(\theta/2)$ where λ is the wave length of the He-Ne laser (λ = 632.8 nm) and θ is the scattering angle. From an initial pressure of $P_i \cong 33$ MPa, for a shallow quench of 0.16 MPa the system enters the metastable region. This is reflected by the continual decay of the scattered light intensities with increasing angle. This behavior is retained at different time scans up to 17.6 seconds shown in the figure. With a small increase in the quench depth to 0.25 MPa the system enters the unstable region and undergoes spinodal decomposition. This is reflected by the maximum in the angular variation of the scattered light intensities. In this system the spinodal ring develops immediately after the quench is imposed, and moves into the observable q range within about 40 ms. These type of experiments are conducted at different polymer concentrations for different quench depths to determine the experimentally accessible spinodal boundary. For some concentrations spinodals are however not experimentally accessible due to the wide metastable gap. Figure 9 shows the binodal and the spinodal boundaries for the PDMS + carbon dioxide system that has been determined by the present technique. Over a concentration range from about 2 to 5.5 %, the crossover from nucleation and growth to spinodal decomposition could be experimentally demonstrated by progressive increase in the quench depth. However in the 0.9 wt % solution, even for a quench depth of 2.4 MPa, spinodal boundary could not be crossed.

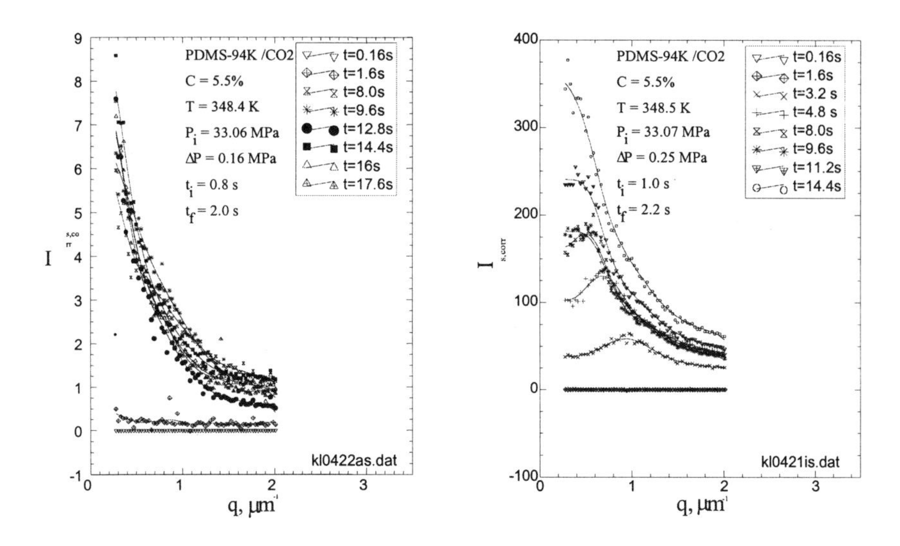

Figure 8. Time evolution of the scattered light intensities as a function of the wave number q = 4π /λ Sin(θ/2) after a pressure quench of 0.16 MPa (left) and 0.25 MPa (right) in a 5.5 % solution of poly (dimethylsiloxane) (PDMS, M_w=94,300) in supercritical carbon dioxide at initial condition of 348 K and 33 MPa.

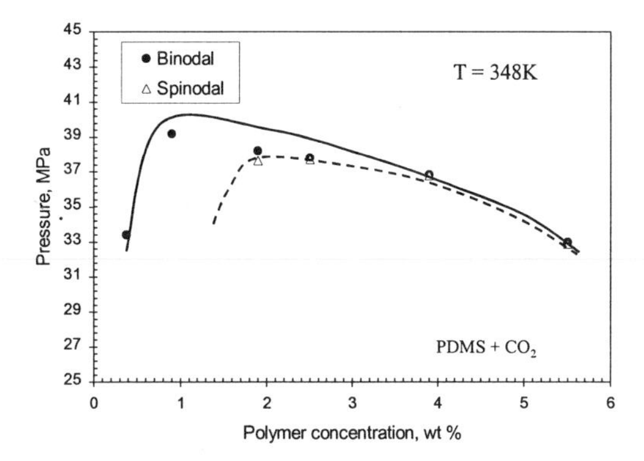

Figure 9. Binodal and experimentally accessible spinodal curves for poly (dimethyl siloxane) [M_w= 94,300; M_w/M_n = 2.99] + carbon dioxide.

As a second example, Figure 10A shows the angular variation and the time evolution of the light scattering patterns in a 5.75 % solutions of polyethylene (M_w = 121,000; PDI = 4.43) in n-pentane at 423 K. The figure shows that the system is undergoing spinodal decomposition for this relatively deep quench of 1.1 MPa. In this system for shallower quenches, phase separation was observed to proceed by nucleation and growth.[9] Figure 10B is the binodal and spinodal envelopes that were determined for this system at 423 K. As in the case of PDMS + carbon dioxide, unstable domains are relatively easy to enter at certain concentrations. This is the concentration range that includes the critical polymer concentration. Figures 9 and 10 show that for these polymer solutions the spinodal and the binodal merge at concentrations higher than the apex of the binodals. This is a manifestation of the broad molecular weight distribution for these polymers. The selection of broad molecular weight distribution was intentional and helpful to demonstrate the shift in the concentration as an added verification of the power of the technique. (Experiments conducted with narrower molecular weight distribution polyethylenes show that the spinodal and the binodal indeed merge closer to the apex of the binodal).[19]

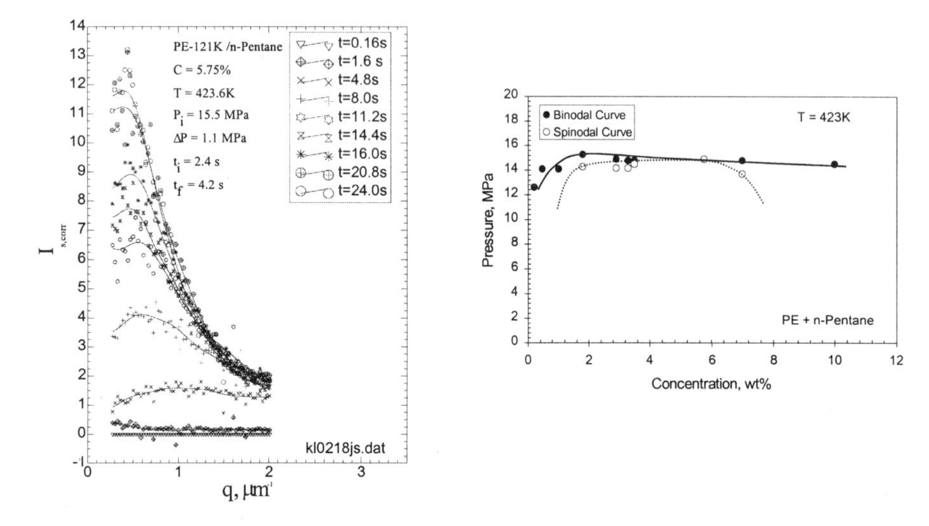

Figure 10. ***A (Left)*** Time evolution of the scattered light intensities as a function of the wave number q $= 4\pi/\lambda \, \mathrm{Sin}(\theta/2)$ after a 1.1 MPa quench in 5.75 % solution of polyethylene (M_w = 121,000; PDI = 4.43) in n-pentane. ***B (right)*** Binodal and the experimentally accessible spinodal boundary for the same system.

It is important to note significance of these experiments in demonstrating that even at concentrations away from the critical polymer concentration the spinodal decomposition regime may be entered by a rapid pressure quench as long as the metastable gap is not too large. It is also important to note that the time evolution of the scattered light intensities in polymer solutions subjected to a pressure quench display fast kinetics. As can be assessed from Figure 8 and 10, the new phase growth progresses rapidly within seconds.

For quenches leading to spinodal decomposition, the location of the scattered light intensity maximum in these experiments does not remain stationary at a fixed angle, but moves to smaller angles (lower q values) in time within the sampling time intervals of the present experimental system. That the intensity maximum is not stationary suggests that the system passes through the early stage of the spinodal decomposition extremely fast (most likely that within microseconds that we cannot capture in the present system) and enters the intermediate and late stages. The scattered light intensity profiles shown in Figures 8 and 10 correspond to the intermediate or late stage of spinodal decomposition and the maximum of scattered light intensity I_m and the peak wave number q_m show power law relationship of the type $q_m \cong t^{-\alpha}$ and $I_m \cong t^{\beta}$ that are typical of later stages of spinodal decomposition.[3,17-19]

The maximum value of the wave number q_m is related[20] to the dominant size scale through $L = 2\pi/q_m$. In these systems the characteristic domain size grows rapidly and reaches a size of about 4 microns within about 1 s after the quench is initiated. This is illustrated in Figure 11 for the case of the PDMS and PE solutions undergoing spinodal decomposition as shown in Figues 8 and 10. The domain size grows from about 4 to 10 microns within 5 s after the pressure quench.

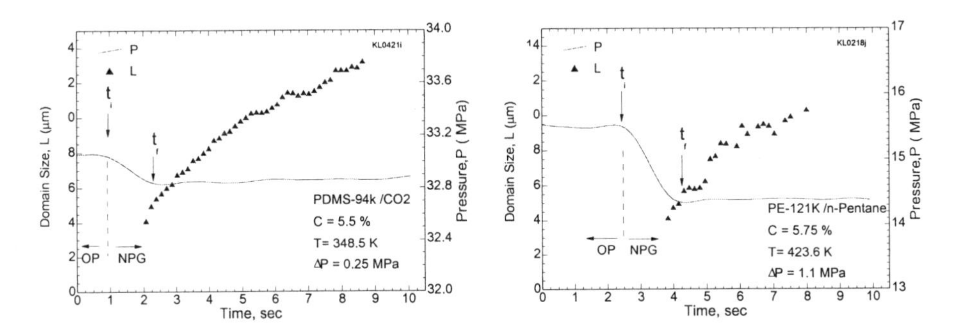

Figure 11. Characteristic domain growth in PDMS + CO_2 (left) and PE + n-pentane (right) solutions undergoing pressure-induced spinodal decomposition. OP = one-phase; NPG = new phase growth. See Figures 8 and 10.

POLYMER BLENDING

We are exploring pressure induced miscibility and phase separation as a technique to produce novel polymer-polymer blends, in particular blends of polymers that are mutually incompatible. In the preceding section on miscibility we have already commented on the solution blending of polyethylene and isotactic polypropylene. A more recent approach that we have been evaluating is based on impregnating a host polymer that swells but does not dissolve in a high-pressure fluid with a solution of a second polymer that dissolves in the same fluid.[21,22] The system is then subjected to a pressure quench upon which the dissolved polymer is in-situ precipitated inside the host polymer. The entrapment is further enhanced with rapid solidification of the host polymer which accompanies the escape of the dissolved fluid from the polymer. Poly(dimethylsiloxane) and polyethylene present a unique polymer pair to explore this type of blending. This is because PDMS, as we have shown in the previous sections dissolves in supercritical carbon dioxide whereas polyethylene does not. High molecular weight PDMS and PE are also known to be highly incompatible.[22] In a recent study, pellets of the polyethylene (M_w = 121,000) sample discussed in the previous sections were impregnated with a 0.1 wt % solution of PDMS (M_w = 94,000), also discussed in the previous sections, in supercritical carbon dioxide. The impregnation was carried out in a view cell loaded with the polymer pellets, PDMS and CO_2 at 130 °C and 57 MPa. The temperature of impregnation was chosen to be close to the crystalline melting temperature of polyethylene so that it can be swollen with carbon dioxide. At these conditions, the translucent appearance of the initially opaque polymer pellet provides a visual verification of the penetration of carbon dioxide into the host polymer matrix. The pressure conditions were chosen to insure complete miscibility for PDMS in carbon dioxide. After equilibration at these conditions, the pressure was reduced rapidly to atmospheric pressures by venting the carbon dioxide from the cell. The depressurization was carried out either at the initial impregnation temperature of 130 °C, or at lower temperature of 50 °C. The pellets were then removed, freeze-fractured in liquid nitrogen, carbon-coated and characterized by Energy Dispersive X-Ray Spectroscopy (EDX) for silicon atom distribution across the cross-section. The microstructural changes were also examined by electron microscopic investigations. Figure 12 shows the silicon count across the cross-section of the pellets. Even though near the outer edges of the pellet a decrease in the silicon count is noted, an average net silicon count of about 5000 is observed throughout the pellet. This overall picture is similar in the pellets depressurized at 130 °C or at 50 °C. Closer analysis of the distribution in the first 100 micron range from the outer edges give indications that silicon count shows an initial decrease but then passes through a localized maximum. This phase localization and unique gradient formation becomes more distinctly observable and highly dependent on the temperature of depressurization when the host polymer is a glassy polymer like polystyrene instead of semicrystalline polymer like polyethylene.[22] This localization has been attributed to the vitrification /crystallization – enhanced entrapment of PDMS while it is carried away from the inner regions of the pellet towards the outer surface during decompression.

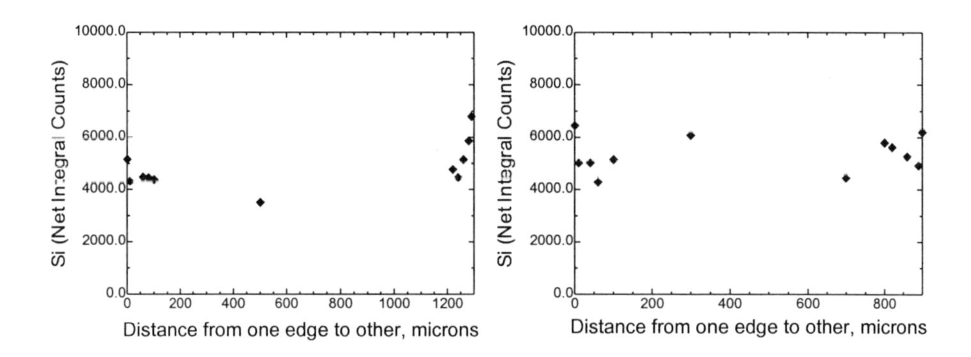

Figure 12. Silicon distribution in polyethylene blended with PDMS by impregnation with 0.1 % solution of PDMS in carbon dioxide at 57 MPa and 130 °C and depressurization at 130 °C (left) and 50 °C (right).

A consequence of pressure-induced phase-separation in polymers swollen with a fluid like supercritical carbon dioxide is the foaming and porous structure formation. Figure 13 shows the microstructure of the polyethylene pellet before and after the impregnation (at 130 °C) with 0.1 % PDMS solution and depressurization (at 50 °C) treatment at a magnification of about X 10,000. While the initial polymer is nonporous, during the process a nano-porous polyethylene has been produced. Fine-tuning of these processes to achieve such unique microstructures and compositional distributions is a new challenge. It requires a balanced understanding of the thermodynamics (miscibility and phase separation), kinetics (time scale of phase growth, diffusional rates) and other physicochemical properties of the fluids and the materials such as the viscosity, glass transition temperature, interfacial tension, melting and crystallization temperatures and their compositional dependence at high pressures.

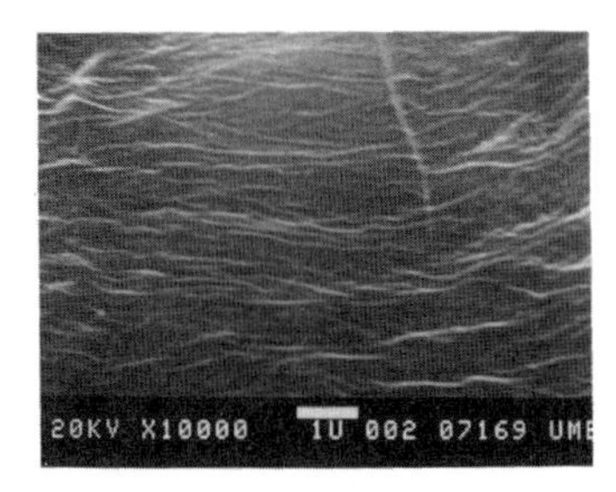

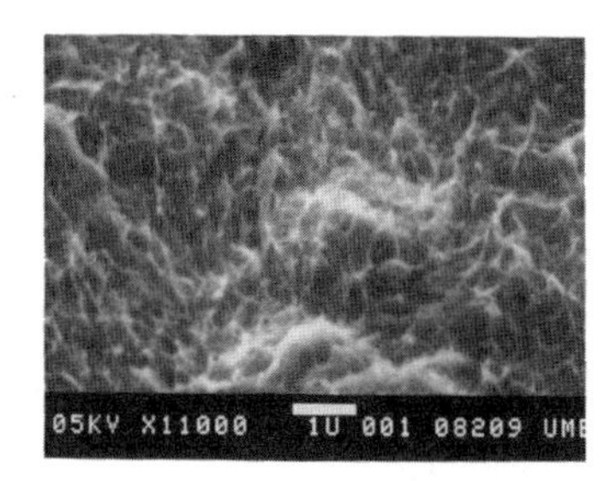

Figure 13. Electron micrographs of polyethylene before (left) and after swelling in 0.1 % solution of PDMS in carbon dioxide at 57 MPa and 130 °C followed by depressurization at 50 °C (right).

References

1. E. Kiran, Polymer miscibility and kinetics of pressure-induced phase separation in near-critical and supercritical fluids, Chapter 6 in *Supercritical Fluids. Fundamentals and Applications*, E. Kiran, P. G. Debenedetti and C. J. Peters, Eds., Kluwer Academic Publishers, Dordrecht, The Netherlands (2000).
2. W. Zhuang and E. Kiran, Kinetics of pressure-induced phase separation (PIPS) from polymer solutions by time-resolved light scattering. Polyethylene + n-pentane, *Polymer*, 39, 2903-2915 (1998).
3. T. Hashimoto, Structure formation in polymer mixtures by spinodal decomposition, in *Current Topics in Polymer Science*, Vol II, R. M. Ottenbrite and L. A. Utracki, Eds., Hanser, New York (1987).
4. L. A. Utracki, Thermodynamics and kinetics of phase separation in *Interpenetrating Polymer Networks*, D. Klemper and L. H. Sperling, Eds., ACS Advances in Chemistry Series, Vol. 239, ACS, Washington, D.C., (1994); pp. 77-123
5. E. Kiran, Polymer formation, modification and processing in or with supercritical fluids, in *Supercritical Fluids. Fundamentals for Applications*, E. Kiran and J. M. H. Levelt Sengers, Eds. Kluwer Academic Publishers, Dordrecht, The Netherlands (1994); pp. 541-588.
6. E. Kiran and W. Zhuang, Miscibility and phase separation of polymers in near- and supercritical fluids, in *Supercritical Fluids- Extraction and Pollution Prevention*, M. A. Abraham and A. K. Sunol, Eds, ACS Symp. Ser. No 670, ACS, Washington, D. C., (1997); pp. 2-36.
7. B. Folie and M. Radosz, Phase equilibria in high-pressure polyethylene technology, *Ind. Eng. Chem. Res.*, 34, 1501-1516 (1995).
8. A. D. Shine, Polymers in supercritical fluids, Chapter 18 in *Physical Properties of Polymers Handbook*, J. E. Mark, Ed., American Institute of Physics, New York (1996).
9. C. F. Kirby and M. A. McHugh, Phase behavior of polymers in supercritical fluid solvents, *Chem. Rev.*, 99, 565-602 (1999).
10. Z. Bayraktar and E. Kiran, Miscibility, phase separation and volumetric properties of solutions of poly(dimethylsiloxane) in supercritical carbon dioxide, *J. Appl. Polym. Sci.*, 75, 1397-1403 (2000).
11. H.-G. Elias, *An Introduction to Polymer Science*, VCH, New York (1997), p.231.

12. W. Zhuang, Miscibility, demixing and kinetics of phase separation of polymers in near and supercritical fluids by light scattering, *Ph. D. Thesis*, University of Maine, Orono, Maine, USA (Thesis Advisor: E. Kiran), 1995.
13. E. Kiran and H. Pöhler, Alternative solvents for cellulose derivatives: Miscibility and density of cellulosic polymers in carbon dioxide + acetone and carbon dioxide + ethanol binary fluid mixtures, *J. Supercrit. Fluids*, 13, 135-147 (1998).
14. E. Kiran, Y. Xiong, and W. Zhuang, Effect of polydispersity on the demixing pressures of polyethylene in near- and supercritical alkanes, *J. Supercrit. Fluids*, 7, 283-287 (1994).
15. H. Pöhler and E. Kiran, Miscibility and phase separation of polymers in supercritical fluids: Polymer-polymer–solvent and polymer-polymer-solvent ternary mixtures, presented at the *AIChE Annual Meeting*, Chicago, Illinois, November 10-15, 1996.
16. Y. Xiong and E. Kiran, High-pressure light scattering apparatus to study pressure-induced phase separation in polymer solutions, *Rev. Sci. Instrum.*, 69, 1463-1471 (1998).
17. Y. Xiong and E. Kiran, Kinetics of pressure-induced phase separation (PIPS) in polystyrene + methylcyclohexane solutions at high pressure, *Polymer*, 41, 3759-3777 (2000).
18. K. Liu and E. Kiran, Kinetics of pressure-induced phase separation (PIPS) in solutions of poly(dimethylsiloxane) in supercritical carbon dioxide: Crossover from nucleation and growth to spinodal decomposition. *J. Supercrit. Fluids*, 16, 59-79 (1999).
19. K. Liu and E. Kiran, Pressure-induced phase separation (PIPS) in polymer solutions: Kinetics of phase separation and crossover from nucleation and growth to spinodal dccomposition in solutions of polyethylene in n-pentane, *Proc. 5th Int. Symp. Supercritical Fluids*, ISSF 2000, April 9-11, 2000, Atlanta, Georgia.
20. T. Hashimoto, Self assembly of polymer blends at phase transition- morphology control by pinning of domain growth, in *Progress in Pacific Polymer Science*, Vol 2., Y. Imanishi, Ed., Springer, Berlin (1992), pp.175-187.
21. Z. Bayraktar and E. Kiran, Microblending of polymers with swelling and impregnation from supercritical fluid solutions. An investigation of the polymer distribution using energy dispersive X-ray spectroscopy, Presented at the *AIChE Annual Meeting*, Dallas, Texas, October 31-November 5, 1999.
22. Z. Bayraktar and E. Kiran, Polymer blending by pressure-induced swelling. Impregnation and phase separation in supercritical fluids. Gradient blending and phase localization of poly(dimethylsiloxane) in polystyrene and polyethylene, submitted for publication in *Macromolecules*.

THE COMPATIBILIZATION OF POLYMER BLENDS WITH LINEAR COPOLYMERS: COMPARISON BETWEEN SIMULATION AND EXPERIMENT

M. D. Dadmun

Chemistry Department
The University of Tennessee
Knoxville, TN 37996-1600

INTRODUCTION

Polymers offer a wide range of material properties, from the flexibility of polyethylene in garbage bags to the high strength of poly(phenylene terephthalamide) in bulletproof jackets. This wide range of polymer properties has led to the widespread use of polymers in many materials applications. However, when a new application arises in industry, often a single polymer can not fulfill all of the material requirements that are associated with that application. In this case, development of a new polymeric material may be needed to fulfill the material property requirements. In these situations, industry has often utilized mixing of two (or more) polymers to develop new materials with targeted properties. By combining two polymers with diverse properties, it is possible to create a new material that retains physical characteristics of both polymers. However, it is also well known that two long chain molecules will rarely mix on a thermodynamic level due to their low entropy of mixing. The resultant two-phase structure will have inferior properties to the initial components, primarily due to the presence of a sharp biphasic interface that does not provide entanglement between the polymers in the separate phases. This lack of entanglement across the interface results in poor transfer of stress, which in turn degrades the macroscopic properties of the mixture.

Due to the importance of the presence of a biphasic interface on the macroscopic properties of a polymer blend, substantial work has been completed towards understanding and improving the interface and thus the macroscopic properties of the mixture. In particular, the effect of adding a copolymer to act as an interfacial modifier has received abundant attention. Much of this work has centered on the ability of a copolymer to strengthen the biphasic interface, lower interfacial tension (to create a finer dispersion), and inhibit coalescence during processing. Each of these mechanisms apparently contributes to the improvement of macroscopic properties of biphasic polymer blends upon addition of a copolymer and the importance of each has been the subject of some debate in the literature.

Computational Studies, Nanotechnology, and Solution Thermodynamics of Polymer Systems
Edited by Dadmun *et al.*, Kluwer Academic/Plenum Publishers, New York, 2000

Specific results that have been published include the work of Balasz et. al. who have completed extensive self-consistent calculations and Monte Carlo simulations to evaluate the effect of adding copolymers to a blend on the interfacial tension of the biphase.[1-9] Additionally, Kramer and others have utilized experimental techniques to determine the mechanism of fracture that occurs at a biphasic interface in the presence or absence of copolymer.[10-27] Macosko, however, has viewed the problem from an engineering perspective and has examined the role of added copolymer on the blend morphology that results from typical processing conditions.[28-31]

Much of the work on copolymers as interfacial modifiers has utilized block copolymers as additives. The role of copolymer molecular weight, composition, and other molecular parameters on the ability of a block copolymer to improve the properties of a biphasic blend is well understood. However, block copolymers are expensive and difficult to synthesize. Therefore, their use as interfacial modifiers in commercial applications has been limited. Other copolymer structures, including random copolymers, may also act as compatibilizers. However, there exist conflicting results regarding the utility of random copolymers as interfacial modifiers.[11,12,13,25]

These results include studies on mixtures of polystyrene (PS) and poly(methyl methacrylate) (PMMA) with random copolymers of PS and PMMA as compatibilizer which show that the random copolymer does not improve the strength of the biphasic interface as well as a block copolymer.[12,13] Alternatively, if the same experiment is completed on mixtures of PS and poly(2-vinyl pyridine) (P2VP) with a PS/P2VP random copolymer, the random copolymer substantially improves the interfacial strength between the two homopolymers[11,25] Additionally, seminal work by Fayt et. al. [32-36] showed that addition of tapered (which can also be thought of as a type of random) copolymers of PS/PE (polyethylene) to a blend of PS and PE improved the elongation and ultimate strength of that blend MORE than the addition of a similar diblock copolymer.
In this paper, Monte Carlo simulation studies will be discussed which provide insight into the underlying factors that affect the ability of a copolymer to strengthen and interface and compatibilize a polymer blend. The interpretation of these results will then be correlated to the experimental evidence that currently exists in the literature. It is expected that the results of this work will provide important fundamental information on the underlying physics that govern the interfacial behavior of copolymers. In turn, this information can be utilized to develop processing schemes by which materials can be efficiently created from polymer mixtures with optimized and tunable properties.

MONTE CARLO TECHNIQUE

The model system[37, 38] in this study consists of 2504 chains of length N=10 confined to a cubic lattice. To simulate an infinite set of chains, the system is approximated as a set of infinitely many identical cells of length L with periodic boundary conditions in all three orthogonal directions (x, y, z). In this study, the only interaction energy is a nearest neighbor monomer-monomer interaction, ε_{A-B}. ε_{A-B} is zero if two neighboring monomers are of different type (A-B) and is negative otherwise. ε_{A-B} applies to any two adjacent monomers, whether a bond connects them or not. Steric interactions are included as excluded volume; simultaneous occupation of a given lattice site by more than one monomer is prohibited. The density of the system is calculated as the fraction of occupied lattice sites, $\rho = N_pN/L^3$, where $N_p = 2504$ is the number of polymer chains present, N=10 is the length of all chains, homopolymers and copolymers, and L=31 is the size of the cubic lattice. In the present study, $\rho = 84.05\%$.

Adding the polymers to the lattice in a completely ordered state creates the initial configuration. One half of the homopolymers are of type A and half are type B. The percentage of copolymer present is varied in the study, though most of the trends that are described below are independent of copolymer concentration. The composition of the copolymer is 50% A and 50% B, while the sequence distribution of the copolymer is parameterized by the parameter Px, which is equal to :

$$Px = P_{AB} / (P_A \times P_B) \tag{1}$$

In this equation, P_{AB} is the percentage of AB dyads in the copolymers, P_A is the percent of A monomers in the copolymer and P_B is the percentage of B monomers in the copolymer. Px is essentially a normalized probability that two neighboring monomers on a copolymer chain are different types.
Once the initial configuration exists, applying a modified reptation technique[39] to the chains creates various chain configurations. In this modified reptation technique, a void on the lattice is chosen at random. A direction from that void is selected randomly. If the end of a chain resides on that lattice point, the polymer chain is reptated into the void and the other end of the chain is vacated. In this way the configuration of the polymer chain changes and the void is moved. The new configuration is accepted according to the Metropolis sampling technique.[40] To ensure there is no bias due to the initially ordered state, the system is equilibrated through 10,000 system configurations before statistics for the system characterization are calculated.

The behavior of the copolymers as a compatibilizer will be analyzed by examining the change in the configuration and volume of the copolymer. The configuration of the copolymer is given by the radii of gyration of the copolymer along each axis, r_gx, r_gy, and rgz.
The radii of gyration of the copolymer are calculated using the following formulae:

$$r_g w = \frac{1}{nch} \sum_{i=1}^{nch} \left[\frac{1}{2N^2} \sum_{l=1}^{N} \sum_{k=1}^{N} \left(w(i,l) - w(i,k) \right)^2 \right] \tag{2}$$

where w = x, y, or z, *nch* is the number of copolymer chains, and $w(i,l)$ is the position on the w axis of the lth monomer on the ith copolymer chain.

To minimize statistical deviation, each point in the following figures is an average of at least 7.5×10^6 Monte Carlo steps per chain in the vicinity of the phase transition and at least 2.0×10^6 Monte Carlo steps per chain far from the phase transition. This simulation was completed on many computers including a DEC Alpha 400/266 in the Chemistry Department at the University of Tennessee, a DEC Alpha 3000/700 at the University of Tennessee Computing Center, and an IBM SP/2 at the Joint Institute of Computational Science. The program was run with vector processing and maximum optimization on all machines.

RESULTS AND DISCUSSION

The purpose of these simulations is to evaluate the importance of copolymer sequence distribution on its ability to compatibilize a biphasic interface. Thus, an understanding of the compatibilization process is necessary to decide how this ability can be analyzed. The added copolymer is thought to act as a polymeric surfactant in that it migrates to the biphasic interface of a phase separated blend and lowers the interfacial tension between the two phases. The copolymer also overcomes an important shortcoming of a phase-separated blend as it strengthens the interface. In a non-compatibilized blend, the biphasic interface is sharp and there is very little entanglement between the two phases, which leads to a very weak interface and poor transfer of stress. The copolymer, however, can strengthen that interface by entangling with both homopolymers. This is visualized most easily for a diblock copolymer where the A block unfurls into the homopolymer A phase and entangles with the homopolymer and the B block does the same thing with homopolymer B. It is also thought that the copolymer acts as a buffer at the biphasic interface to inhibit droplet coalescence, which in turn results in a finer dispersion of the minor phase in the major phase.

Thus, to act as a useful compatibilizer in a phase separated system, the copolymer must expand at the interface as it entangles with the homopolymer phases. The copolymer must also take up substantial volume to act as a buffer and be able to inhibit droplet coalescence. Thus, the temperature dependence of the volume of the copolymer at the biphasic interface can be utilized as an assessment of the ability of a copolymer to act as a compatibilizer. If the copolymer entangles with the homopolymers, the volume will swell as the temperature is lowered and the system becomes more phase-separated. However, if the copolymer does not entangle with the homopolymers at the interface, it will be trapped between the two homopolymer phases and collapse as the temperature is lowered.

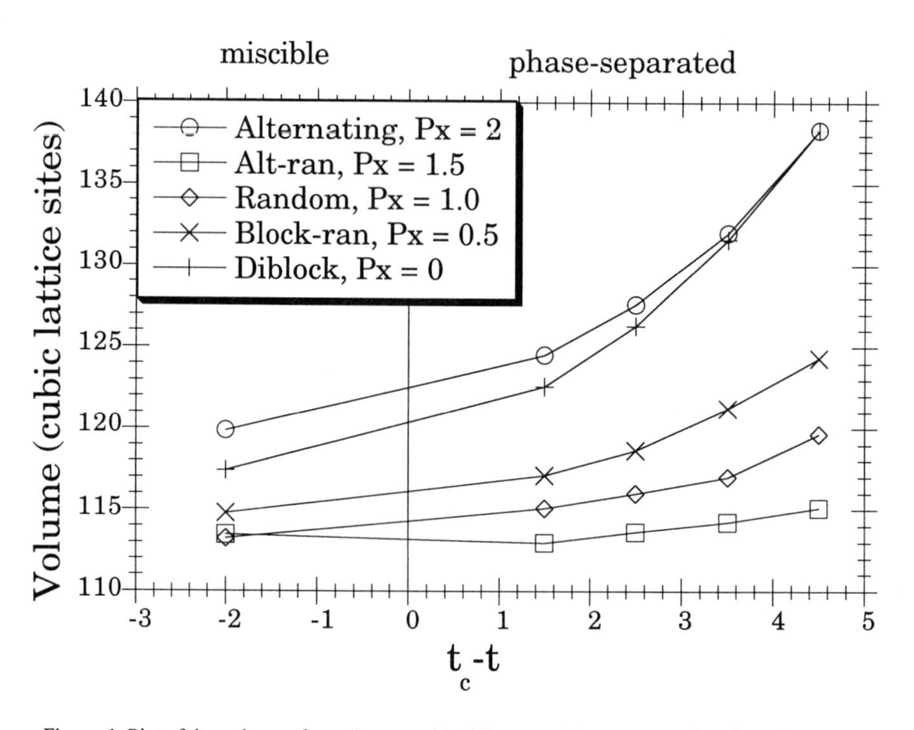

Figure 1. Plot of the volume of copolymers with different architectures as a function of temperature

The volume of the copolymers examined in this study as a function of temperature is shown in Figure 1 for a system with 7.5% copolymer. Blends with other percent copolymers show similar trends. The volumes are calculated from the three radii of gyration as $V = 5^{3/2} \times r_gx \times \cdot r_gy \times r_gz$. the $5^{3/2}$ factor is a result of the relationship between the radius of gyration, moment of inertia and principal axes of an ellipse, which the chain approximates in these conditions.

Inspection of this figure shows that both the alternating and diblock copolymers exhibit a steady and significant increase in the volume of the copolymer as the system becomes more phase-separated. One explanation for this response is that the copolymer is being swollen by its interaction with the homopolymer, suggesting that the copolymer is entangled with the homopolymers in these systems, and will, thus, be an effective compatibilizer. However, examination of the 'random' copolymers (Px = 0.5, 1.0, 1.5) shows an interesting trend. The volume of the alt-ran copolymer (Px = 0.5) does not change much as the system undergoes phase-separation, and actually decreases in volume slightly as the system begins to phase separate. The statistically random copolymer (Px = 0) is slightly better in that it shows a constant increase in size with phase separation, suggesting some interaction with the homopolymers, however the increase in volume (and therefore interaction with homopolymer) is modest. The blocky random copolymer, however shows a steady, significant increase in volume as the system becomes more phase separated.

Thus, from this data, it appears that both the alternating and diblock copolymers are the most efficient at compatibilizing polymer blends. Unfortunately, these two copolymers are also the most difficult to realize and therefore, this is of little use from a commercial standpoint. It is interesting that the alternating copolymer may rival the diblock copolymer as a compatibilizer and this possibility is currently under investigation in our laboratory. Within the random copolymers (Px = 0.5, 1.0, and 1.5), these results suggest that the blocky structure is much better at interacting with the homopolymers than the alt-ran structure, and thus should be a more effective interfacial modifier.

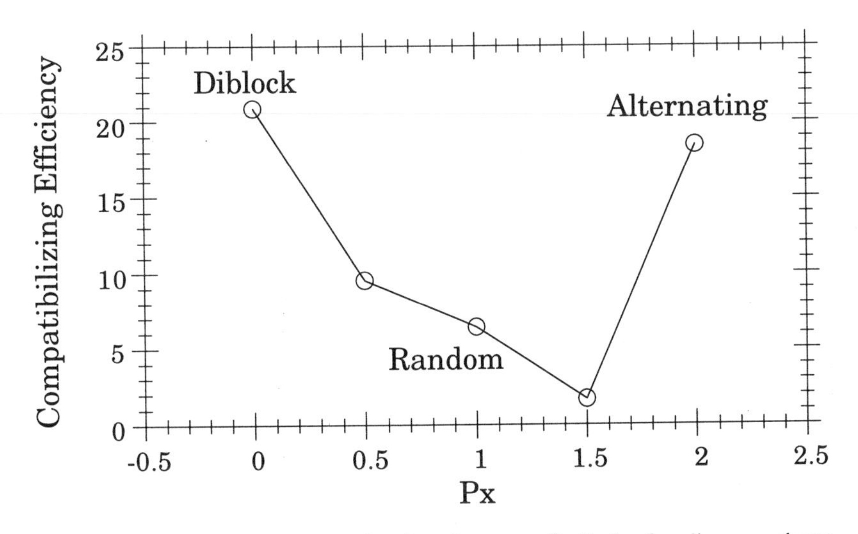

Figure 2. Plot of compatibilizing efficiency as a function of sequence distribution for a linear copolymer.

The change in volume of the copolymer as the system goes from a miscible to an immiscible state has been utilized as a qualitative measure of its ability to compatibilize a biphasic blend. This can be quantified by using the difference between the volume of the copolymer in the miscible system and the volume of the copolymer in the phase separated system at its deepest quench as a measure of the effectiveness of the copolymer as a compatibilizer. This value is plotted vs. the sequence distribution, Px in Figure 2. This data quantifies the trend that is described above; the alternating and diblock copolymers are the best compatibilizers, however within the random structures, the more blocky structures is a more effective interfacial modifier than the statistically random copolymer which is more effective than an alternating-random structure.

Comparison to Experimental Results

It is interesting to compare these calculations to recent experimental results that have examined the ability of 'random' copolymers to strengthen a biphasic interface. To make this comparison, an experimental parameter must be found that can correlate the sequence distribution of the experimental copolymers to the simulation parameter Px. Px is essentially a measure of the percentage of neighboring monomers along the copolymer chain that are of different type. The reactivity ratios of the two monomers can also be utilized to ascertain the propensity of two different monomers to be bonded together along the copolymer chain. Recall that r_1, the reactivity ratio of monomer A, equals k_{11} / k_{12} ; the ratio of reactivity of an A monomer with another A monomer to it's reactivity with a B monomer. Therefore, if $r_1 < 1$, there is more probability that an A monomer will react with a B monomer to continue chain growth, whereas if $r_1 > 1$, there is a greater probability that an A monomer will react with another A monomer than it will react with a B monomer. Thus the copolymer that is formed when $r_1 < 1$ will be more alternating in nature, whereas, if $r_1 > 1$, the copolymer will be more blocky in nature. Similar arguments can be made regarding r_2. Therefore, conceptually, the reactivity ratios define similar characteristics regarding the sequence distribution of a copolymer to the simulation parameter Px; they both relate the propensity to form bonds between dissimilar monomers along the copolymer chain. Therefore, the reactivity ratios of copolymers used in the experimental studies will be utilized to determine the corresponding Px value of the copolymers.

Modeling the copolymerization on a computer completes this correlation between Px and the reactivity ratio. In this model, one million chains that are 2,000 monomers long are synthesized. The composition of the monomer pool is initially set to 50% of each monomer. 1000 chains are grown simultaneously and then repeated until 1×10^6 chains are created. This is meant to account for the broad initiation times that may occur in free radical polymerizations. The reactivity ratios of the monomer pairs as well as the composition of the remaining monomer pool control the evolution of the sequence and composition distributions of the copolymer chain. The probability, P_{AA}, that a monomer **A** will add to the growing chain that ends in an **A**· monomer radical is

$$P_{AA} = \frac{r_1[A]}{r_1[A]+[B]} \tag{3}$$

where r_1 is the reactivity ratio for the A monomer, [A] is the concentration of monomer A in the monomer pool, and [B] is the concentration of monomer B in the monomer pool. Similar equations for P_{AB} (= 1- P_{AA}), P_{BB}, and P_{BA} (= 1 - P_{BB}) can be derived. As a copolymer chain grows, these probabilities are utilized to statistically determine the identity of the next monomer that is added to the growing polymer chain. The final copolymers are then analyzed to determine the corresponding Px values.

The experimental results that will be examined consist of studies that look at the ability of a 'random' copolymer to improve the properties of mixtures of the two homopolymers relative to the ability of a diblock copolymer. The three different systems that are examined include copolymers of poly(styrene-co-methyl methacrylate) (S/MMA), poly(styrene-co-2-vinyl pyridine) (S/2VP), and poly(styrene-co-ethylene) (S/E) in mixtures of the two homopolymers. The experiments that have been utilized to examine the ability of the copolymer to strengthen a polymer blend include the examination of the tensile properties of the compatibilized blend and the determination of the interfacial strength between the two homopolymers using asymmetric double cantilever beam (ADCB) experiments.

The first set of experiments that will be considered has examined the ability of random copolymers of styrene and methyl methacrylate to improve the interfacial strength between polystyrene and poly(methyl methacrylate). [12,13,18] Using the asymmetric double cantilever beam technique, the researchers have found that a diblock copolymer (50/50 composition, Mw = 282,000) creates an interface with strength of 400 J/m^2. When utilizing a random copolymer however, it was found that the strongest interface (70% styrene, Mw = 250,000) had strength of ca. 80 J/m^2. The bare interface between the two homopolymers in the absence of a copolymer had strength of approximately 6 J/m^2. Thus, the random copolymer does strengthen the interface, however not as well as the diblock copolymer. It is interesting to note that Macosko and co-workers have found that random copolymers of styrene and methyl methacrylate do not inhibit static droplet coalescence an applied shear flow, and thus do not function as an effective compatibilizer in that respect.

However, for a copolymer of styrene and 2-vinyl pyridine, similar experiments show a slightly different pattern.[11, 25] Using ADCB, Kramer and coworkers have found that a diblock copolymer (50/50 composition, Mw = 170,000) creates an interface with strength of 70 J/m^2, whereas a random copolymer (50/50 composition, Mw = 700,000) was able to produce an interface with strength of 140 J/m^2. Thus, for this system, the random copolymer is a better interfacial modifier than the diblock copolymer.

The last system that will be compared is that of polystyrene and polyethylene. In these studies,[35] the tensile properties of the blend with an added copolymer were examined. These studies showed that the addition of 10% diblock copolymer (50/50 composition, Mw = 80,000) to a blend that is 80% polystyrene produces an ultimate strength of 33 MPa, whereas the addition of the same amount of a random copolymer (67% polyethylene, Mw = 80,000) produces a blend with an ultimate strength of 41 MPa. Thus, in this system, the random copolymer again creates a superior system to the diblock copolymer.

Interestingly, these results may be explained by a careful examination of the sequence distributions of experimental copolymers. Due to the reactivity ratios of the monomers, the PS/PMMA random copolymer (r_1 = 0.46, r_2 = 0.52) produces a copolymer with a Px value of is 1.28 and is thus alternating in nature, whereas, both PS/P2VP (r_1=0.5, r_2 = 1.3; Px = 0.92) and PS/PE (r_1=0.78, r_2 = 1.39, Px = 0.89) random copolymers are more blocky in nature. Thus, the 'blocky' random copolymer strengthens the interface while the alternating-random copolymer does not. Unfortunately, in all of the experimental studies mentioned above there exist other parameters besides sequence distribution that could influence the correlation of these experimental results to the Monte Carlo work. Thus, further experiments are currently underway in our laboratory to more carefully correlate the influence of copolymer sequence distribution to its ability to modify the biphasic interface in a polymer blend.

Nonetheless, Monte Carlo results suggest that in order for a copolymer to be an effective compatibilizer in polymer blends, the copolymer must be blocky in nature. This interpretation of the simulation results are in agreement with the experimental results which show that the "random" copolymers that are better than the diblock copolymer at compatibilizing a blend are those that are blockier than a statistical random copolymer. The copolymers that do not behave as effectively as a compatibilizer are more alternating than a statistically random copolymer.

CONCLUSION

MC simulation allows careful control of molecular level parameters to allow the examination of the importance of these parameters. An example of one such parameter in copolymers is sequence distribution. MC simulation results have been presented which suggest that 'random' copolymers that are blockier provide improved strength of the biphasic interface than alternating 'random' copolymers. These simulation results also explain seemingly contradictory experimental results. Thus, blocky copolymers are a promising material to provide compatibilization in a phase separated blend as they have been shown to expand at interface in both the parallel and perpendicular directions to allow efficient entanglement with homopolymers. Additionally, they do not microphase separate as easily as diblock copolymers and thus can be more readily driven to the interface.

ACKNOWLEDGEMENTS

The author would like to thank The National Science Foundation, Division of Materials Research (CAREER-DMR-9702313) and 3M Corporation (an Untenured Faculty Grant) for financial support of this research.

REFERENCES

1) R. Isreals, D. Jasnow, A.C. Balazs, L. Guo, J. Sokolov, M.J. Rafailovich, J. Chem. Phys, **102**, 8149 (1995).

2) Y. Lyatskaya, D. Gersappe, A.C. Balazs, Macromolecules, **28**, 6278 (1995).

3) Y. Lyatskaya, S.H. Jacobson, A.C. Balazs, Macromolecules, **29**, 1059 (1996).

4) Y. Lyatskaya, A.C. Balazs, Macromolecules, **29**, 7581 (1996).

5) Y. Lyatskaya, D. Gersappe, N.A. Gross, A.C. Balazs, J. Chem. Phys., **100**, 1449 (1996).

6) G.T. Pickett, D. Jasnow, A.C. Balazs, Phys. Rev. Lett., **77**, 671 (1996).

7) D. Gersappe, D. Irvine, A.C. Balazs, L. Guo, M. Rafailovich, J. Sokolov, S. Schwarz, D. Peiffer, Science, **265**, 1072 (1994).

8) D. Gersappe, A.C. Balazs, Phys. Rev. E, **52**, 5061 (1995).

9) C. Yeung, A.C. Balazs, D. Jasnow, Macromolecules, **25**, 1357 (1992).

10) H. Brown, Ann. Rev. Mat. Sci., **21**, 463 (1991).

11) C.-A. Dai, B.J. Dair, K.H. Dai, C.K. Ober, E.J. Kramer, C.-Y. Hui, L.W. Jelinski, Phys. Rev. Lett, **73**, 2472 (1994).

12) M. Sikka, N.N. Pellegrini, E.A. Schmitt, K.I. Winey, Macromolecules, **30**, 445 (1997).

13) R. Kulasekere, H. Kaiser, J.F. Ankner, T.P. Russell, H.R. Brown, C.J. Hawker, A.M. Mayes, Macromolecules, **29**, 5493 (1996).

14) C. Creton, E.J. Kramer, C.-Y. Hui, H.R. Brown, Macromolecules, **25**, 3075 (1992).

15) J. Washiyama, E.J. Kramer, C.F. Creton, C.-H. Hui, Macromolecules, **27**, 2019 (1994).

16) Y. Lee, K. Char, Macromolecules, **27**, 2603 (1994).

17) E.J. Kramer, L.J. Norton, C.-A. Dai, Y. Sha, C.-Y. Hui Farad. Disc., **98**, 31 (1994).

18) H.R. Brown, K. Char, V.R. Deline, P.F. Green, Macromolecules, **26**, 4155 (1993).

19) H. Brown, J. Mater. Sci., **25**, 2791 (1990).

20) C. Creton, H.R. Brown, V.R Deline, Macromolecules, **27**, 1774 (1994).

21) Y. Sha, C.-Y. Hui, A. Ruina, E.J. Kramer, Macromolecules, **28**, 2450 (1995).

22) H.R. Brown, Macromolecules, **22**, 2859 (1989).

23) K. Char, H.R. Brown, V.R. Deline, Macromolecules, **26**, 4164 (1993).

24) C. Creton, E.J. Kramer, G. Hadziioannou, Macromolecules, **24**, 1864 (1991).

25) C.-A. Dai, C.O. Osuji, K.D. Jandt, B.J. Dair, C.K. Ober, E.J. Kramer, C.-Y. Hui, Macromolecules, **30**, 6727 (1997).

26) K.R. Shull, E.J. Kramer, G. Hadziioannu, W. Tang, Macromolecules, **23**, 4780 (1990).

27) K. H. Dai, E.J. Kramer, K. R. Shull, Macromolecules, **25**, 220 (1992).

28) M.S. Lee, T.P. Lodge, C.W. Macosko, J. Poly. Sci.: Poly. Phys., **35**, 2835 (1997).

29) U. Sundararaj, C.W. Macosko, Macromolecules, **28**, 2647 (1995).

30) C.W. Macosko, P. Guegan, A. Khandpur, A. Nakayama, P. Marechal, T. Inoue Macromolecules, **29**, 5590 (1996).

31) P. Guegan, C.W. Macosko, T. Ishizone, A. Hirao, S. Nakahama, Macromolecules, **27**, 4993 (1994).

32) R. Fayt, R. Jérôme, Ph. Teyssié, J. Poly. Sci.: Poly. Lett. Ed., **19**, 79 (1981).

33) R. Fayt, R. Jérôme, Ph. Teyssié, J. Poly. Sci.: Poly. Phys. Ed., **19**, 1269 (1981).

34) R. Fayt, R. Jérôme, Ph. Teyssié, J. Poly. Sci.: Poly. Phys. Ed., **20**, 2209 (1982).

35) R. Fayt, R. Jérôme, Ph. Teyssié, J. Poly. Sci.: Poly. Phys. Ed., **27**, 775 (1989).

36) R. Fayt, R. Jérôme, Ph. Teyssié, Makro. Chem., **187**, 837 (1986).

37) M.D. Dadmun, Mat. Res. Soc. Symp. Ser. **461**, 123 (1997).

38) M.D. Dadmun, Macromolecules **29**, 3868 (1996).

39) A. Baumgartner, J. Chem. Phys. **1986**, *84*, 1905.

40) N. Metropolis, A.N. Rosenbluth, M.N. Rosenbluth, A.H. Teller, E. Teller, J. Chem. Phys. **1953**, *21*, 1087.

NANOSCALE OPTICAL PROBES OF POLYMER DYNAMICS IN ULTRASMALL VOLUMES

M. D. Barnes,[1] J. V. Ford,[1] K. Fukui,[1] B. G. Sumpter,[1] D. W. Noid,[1] and J.U. Otaigbe[2]
[1]Laser Spectroscopy and Microinstrumentation Group and
Polymer Chemistry Group
Chemical and Analytical Sciences Division, Mail Stop 6142
Oak Ridge National Laboratory
Oak Ridge, Tennessee 37831
[2]Dept. of Materials Science and Engineering
Iowa State University
Ames, Iowa 50011

INTRODUCTION

We describe a new method for production and characterization of polymer and polymer-composite particles from solution using microdroplet techniques. In particular, 2-dimensional optical diffraction – an experimental tool that has long been used for size characterization of liquid droplets – is presented as a sensitive probe of phase separation behavior and particle dynamics in polymer-blend microparticles generated from microdroplets of dilute polymer solution. As we discuss further, this technique is sensitive to the presence of sub-domains with dimensions on the order of 30 nm, thus capable of providing information on material homogeneity on a (macro)molecular length scale. Under conditions of rapid solvent evaporation (i.e., very small droplets) and relatively low polymer mobility, homogeneous particles can be formed using different polymers that ordinarily undergo phase-separation in bulk preparations. This important result paves the way for producing new materials from polymer alloys with tunable material properties.

Over the last several years, an enormous amount of experimental and theoretical effort has been focused on multi-component polymer systems as a means for producing new materials on the micron and nanometer scale with specifically tailored material, electrical and optical properties. Composite polymer particles, or polymer alloys, with specifically tailored properties could find many novel uses in such fields as electro-optic

Computational Studies, Nanotechnology, and Solution Thermodynamics of Polymer Systems
Edited by Dadmun *et al.*, Kluwer Academic/Plenum Publishers, New York, 2000

and luminescent devices,[1,2] conducting materials,[3] and hybrid inorganic-organic polymer alloys.[4] A significant barrier to producing many commercially and scientifically relevant homogeneous polymer blends, [5,6,7, 8] however, is the problem of phase separation from bulk-immiscible components in solution that has been studied in detail by several different groups.[9,10,11] The route typically taken in trying to form homogeneous blends of immiscible polymers is to use compatibilizers to reduce interfacial tension. Recently, a number of different groups have examined phase-separation in copolymer systems to achieve ordered meso- and micro-phase separated structures with a rich variety of morphologies.[12,13] For solvent-cast composites, phase separation and related morphologies depend on the time scale for solvent evaporation relative to molecular organization.

Our interest is in using small droplets (5 - 10 μm diam.) of dilute mixed-polymer solution to form homogeneous polymer composites without compatibilizers as a possible route to new materials with tunable properties. Over the last several years, advances in microdroplet production technology for work in single-molecule detection and spectroscopy in droplet streams has resulted in generation of droplets as small as 2 – 3 microns with a size dispersity of better than 1%. In the context of polymer particle generation, droplet techniques are attractive since particles of essentially arbitrary size (down to the single polymer molecule limit) can be produced by adjusting the size of the droplet of polymer solution, or the weight fraction of the polymer in solution. While droplet production in the size range of 20 – 30 microns (diameter) is more or less routine (several different on-demand droplet generators are now available commercially), generation of droplets smaller than 10 microns remains non-trivial – especially under the added constraint of high monodispersity. Small droplets (< 10 μm) are especially attractive as a means for producing multi-component polymer-blend and polymer-composite particles from solution since solvent evaporation can be made to occur on a millisecond time scale, thus inhibiting phase-separation in these systems.

The primary condition for suppression of phase separation in these systems is that solvent evaporation must occur on a time scale that is fast compared to self-organization times of the polymers. This implies time scales for particle drying on the order of a few milliseconds implying droplet sizes 10 μm (depending on solvent, droplet environment, etc). We have shown recently that a microdroplet approach can be used to form homogeneous composites of co-dissolved bulk-immiscible polymers[14] using instrumentation developed in our laboratory for probing single fluorescent molecules in droplet streams.[15,16] In addition to a new route to forming nanoscale polymer composites, a microparticle format offers a new tool for studying multi-component polymer blend systems confined to femtoliter and attoliter volumes where high surface area-to-volume ratios play a significant role in phase separation dynamics.

Here we describe in some detail the basis of optical diffraction in spherical dielectric particles as a probe of material homogeneity in polymer composites, and discuss limitations on domain size (in multi-phase composites), and dielectric constant. We show how this measurement technique can be used to recover information on drying kinetics, inter-polymer dynamics, and material properties. In the following chapter, we describe some results of detailed molecular dynamics modeling that can be used to connect experimental observables with microscopic dynamics within the particle as well as to suggest future experiments. The organization of this chapter on synthesis and characterization of polymer and polymer-composite particles is as follows: First we describe our instrumentation for production, manipulation, and characterization of polymer particles from microdroplets of solution. Next we summarize some of the important results. Finally, we discuss some exciting possible future directions and applications

EXPERIMENTAL

Diffraction-Based Probes of Material Homogeneity in Polymer-Composite Microparticles

Our primary experimental tool for probing phase-separation behavior and material homogeneity in polymer composites is essentially an interferometric technique that has been used for a number of years as a method for sizing liquid microdroplets (size range ≈ 2 – 20 microns. Recently, we have shown that this measurement technique can be used to recover information on drying kinetics, inter-polymer dynamics, and material properties such as dielectric constant.[17] The basis of the technique involves illumination of a dielectric sphere with a plane-polarized laser to produce an inhomogeneous electric field intensity distribution, or grating, within the particle that results from interference between refracted and totally-internally-reflected waves within the particle. The angular spacing between intensity maxima, as well as the intensity envelope is a highly sensitive function of particle size, and refractive index (both real and imaginary parts). Unlike conventional microscopy approaches with diffraction-limited ($\approx \lambda/2$) spatial resolution, two-dimensional diffraction (or, angle-resolved scattering) is sensitive to material homogeneity on a length scale of $\approx \lambda/20$ or about 20 – 30 nm for optical wavelengths. This dimension is comparable to single-molecule radii of gyration for relative large molecular weight (> 100 k) polymers, and thus provides molecular scale "resolution" of material homogeneity in ultrasmall volumes (≈ 1 – 100 femtoliters).

Light scattering from micron-sized spherical dielectric droplets or particles[18] has been used for a number of years as a method of sizing[19,20,21] and analysis of various physical and chemical properties.[22,23] While various light scattering techniques from spherical droplets have been very well characterized for particle sizing and refractive index determination.[24,25,26,27] use of 1- and 2-dimensional angle-resolved elastic scattering has very recently begun to be utilized as a tool for characterizing in *situ* polymerization in microdroplets,[28,29] and probing multi-phase[30,31] and homogeneous[14] composite particles. As shown in Ref. 14, the fringe contrast (and intensity fluctuations along an individual fringe) is also very sensitive to material homogeneity on a length scale of $\approx \lambda/20$ or about 20 – 30 nm for optical wavelengths. In our experimental configuration, this intensity grating is projected in the far-field using (f/1.5) collimating optics and detected with a CCD camera.[32] The scattering angle (center angle and width) is established by means of an external calibration, and is used for high-precision Mie analysis of one-dimensional diffraction data.

For optical diffraction studies, individual particles were studied using droplet levitation techniques. Details of the apparatus and CCD calibration procedure are described in Ref. 32. The nominal scattering angle was 90 degrees with respect to the direction of propagation of the vertically polarized HeNe laser, and the useable full plane angle (defined by the f/1.5 achromatic objective) was 35 degrees. The CCD (SpectraSource Instruments) was thermoelectrically cooled and digitized at 16 bits. Details of the droplet generator used are described in Ref. 16. Aqueous solutions were handled by

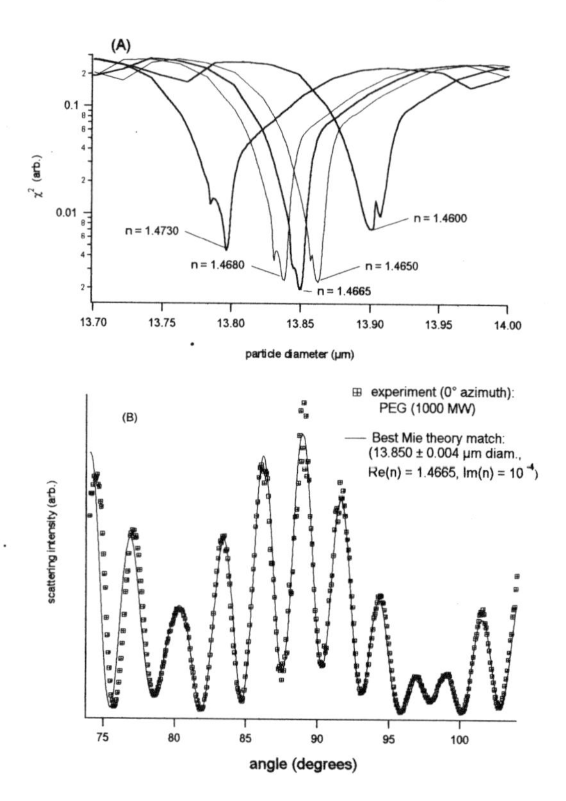

Figure 1. Two-dimensional slices of 4-dimensional error surface (varying Re(n)) for Mie theory match to diffraction from PEG particle (Im(n) fixed). The lower trace shows the best match to the experimental scattering data.

simply loading the Pyrex tip by vacuum aspiration and re-installing into the generator. For the work done on co-dissolved polymers in tetrahydrofuran (THF), the entire droplet generator chamber and ballast reservoir were backfilled with THF, and the tip was loaded with the polymer solution of interest.

RESULTS

Figure 1 shows a family of 2-D slices (Re(n) varied and Im(n) fixed) of a typical error surface obtained from data acquired for a PEG particle. The three independent factors that define the scattering pattern – size, Re(n), and Im(n) – are systematically varied to find the best possible match to the experimental data by locating the minimum in a 4-dimensional error function. From exhaustive analysis of these error surfaces from many different sized particles, we find absolute size uncertainties to be between 2 and 5 nanometers, and the uncertainty in Re(n) to be between 10^{-3} and 5 x 10^{-4}. For materials with a low molar absorptivity (typical of most dielectric liquids), Im(n) is correspondingly small – on the order of 10^{-5} to 10^{-7}. At these values, there is very little (if any) effect on the match to data by varying Im(n). For many polymers (polyvinyl chlorides for example) however, this is not the case and Im(n) can be as large as 10^{-3}. At this order of magnitude, Im(n) does indeed influence the Mie analysis of the data.

A key issue in forming homogeneous composites from co-dissolved bulk-immiscible polymers from solution is that the droplet evaporation rate must be fast compared with the polymer self-organization time scale. Since the time scale for solvent evaporation is proportional to $1/(r^{3/2})$, r is the droplet radius, the most obvious way to satisfy this condition is to make droplets smaller – an option which is nontrivial experimentally.[15] Alternatively, one can modify the atmosphere around the droplet (temperature, different bath gases, etc.) to accelerate particle drying.[16] As part of our effort in developing single-molecule isolation and manipulation methodologies in droplet streams, we have been able to produce (water) droplets as small as 2 – 3 microns in diameter with 1% size dispersity. For a 3-µm (nom.) droplet, approximately 95% of the initial droplet volume is lost due to evaporation in about 5 ms. The diffusion coefficient of a free polymer molecule in solution with molecular weight of say, 10^4, is on the order of 10^{-9} cm^2/sec. In this case, diffusional motion of the polymer center-of-mass is negligible (<r> 30 nm) on the time scale of solvent evaporation.

Material Properties and Particle Dynamics

Another important aspect of our approach is the ability to obtain very precise information on the refractive index which provides a way of characterizing material properties and particle dynamics that is not possible with conventional microscopy techniques. One can, for example, directly measure kinetics of particle drying by monitoring the refractive index change as the solvent evaporates. In general, we find that the particle drying kinetics have two distinct solvent evaporation regimes. When the droplet is first ejected, the size decreases rapidly with most of the solvent evaporation taking place within the first 10 to 100 milliseconds. This is followed by a much slower ("wet particle") evaporation regime where the particle continues to lose solvent on a time scale of several minutes (dependent on size). Interestingly, in the slow-evaporation regime, we generally observe that the volume change expected from solvent loss (assuming an ideal solution) is *not* accompanied by the expected corresponding change in particle size.

Put another way, we observe a much smaller change in size than would be expected from the change in refractive index (solvent loss) if the particle were able to respond to solvent removal in the same way as an ideal solution. This suggests that, in most cases, the polymers form a semi-rigid matrix through which trapped residual solvent escapes by diffusion.

We examined two bulk-immiscible polymers (polystyrene and polyvinyl chloride) that are soluble in THF.[14] We have not characterized evaporation rates of pure THF droplets, but estimate on the basis of vapor pressure differences relative to water, that evaporation is roughly a factor of 5 more rapid (for a given droplet size). To provide some kind of relative time scale, a water droplet with an initial diameter of 10-μm will evaporate to 1 μm diameter in about 5 ms in a dry Argon atmosphere.[16]

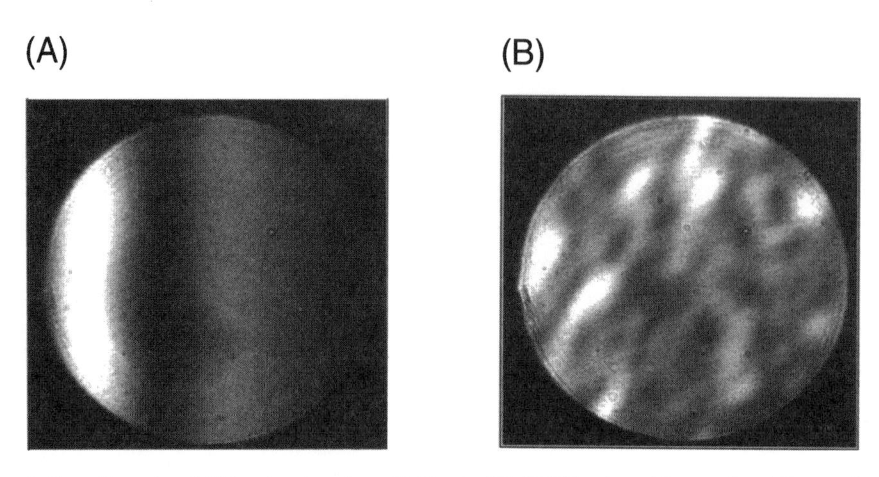

Figure 2. Two-dimensional diffraction data from 50:50 w/w PVC/PS blend particles produced from an 8 μm diameter droplet (left), and a 35 μm diameter droplet. (right).

Figure 2 shows examples of PVC/PS particles formed from an (a) 8 μm diameter droplet, and (b) a 35 μm diameter droplet of dilute (1% total polymer weight fraction) PVC/PS/THF solution. The size threshold (for this system) for producing homogeneous particles is about 10 μm. As shown in Figure 2-b, for larger droplets that have correspondingly longer drying times, we observed that the particles were inhomogeneous as evidenced by the absence of well defined (vertical) diffraction fringes. There is compelling evidence for material homogeneity at a molecular length scale for the particle represented in Figure 2-a: fringe uniformity, quantitative agreement with Mie calculations, and a refractive index (related to material dielectric constant) that is intermediate between the two pure materials.

Figure 3 shows time-resolved results of particle size and refractive index for a 2.4 μm PVC/PS (50:50 w/w) blend particle. Note that the particle continues to "dry" on a time-scale of several minutes, but remains homogeneous throughout the measurement sequence. The nominal index of pure THF is 1.41 and the measured (steady state) final index of the particle is 1.527. These limiting values suggest that, for the first data point in

Figure 3, the particle is 22% by volume THF. Assuming that the last data point represents a fully dry particle (in good agreement with estimates based on refractive indices for pure PS and PVC), one would have expected a 60% *decrease* in particle size accompanying the loss of residual THF. This is clearly not observed, indicating that the particle has formed a fairly rigid matrix that compresses only slightly (3%) during the remainder of the observation time.

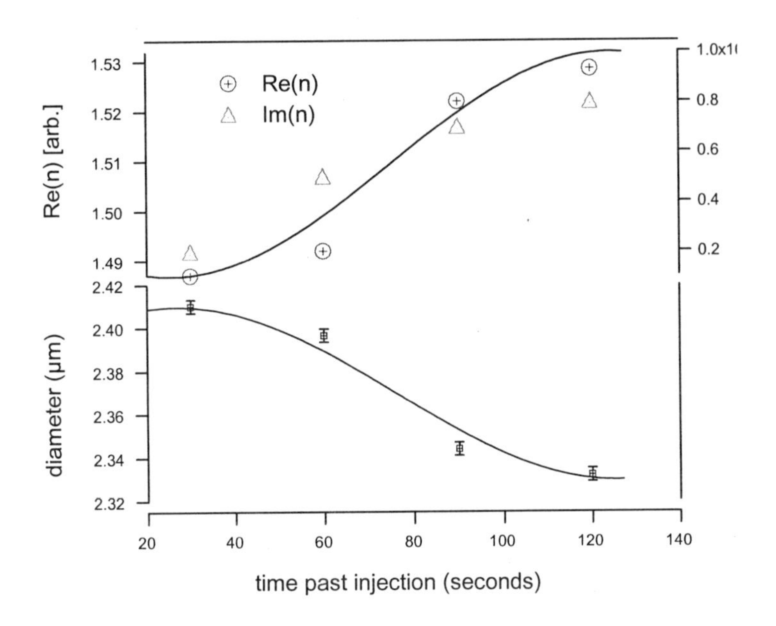

Figure 3. Particle size and refractive index (real and imaginary parts) for a PVC/PS composite particle (50:50 w/w) as a function of time past injection.

The data shown in Figure 3 also illustrates the 'tunable' nature of a material property in PVC/PS composite particles – namely the dielectric constant manifested in the refractive index. Both Re(n) and Im(n) for the polymer-blend microparticles are intermediate between the values determined for pure single-component particles (PVC: Re(n) = 1.4780, Im(n) = 10^{-3} ; PS: Re(n) = 1.5908, Im(n) = 2×10^{-5}) and can be controlled by adjusting the weight fractions of polymers. Interestingly, the measured refractive index for composite particles are very close to estimates obtained from a simple mass-weighted average of the two species.

Domain Size 'Resolution' in 2-D Diffraction

We have recently shown that the presence of phase-separated structures in *heterogeneous* polymer-blend microparticles can be indicated qualitatively by a distortion in the two-dimensional diffraction pattern. The origin of fringe distortion from a multi-phase composite particle can be understood as a result of refraction at the boundary between sub-domains of different polymers, which typically exhibit significant differences in refractive index. Thus, the presence of separate sub-domains introduces cumulative optical phase shifts and refraction resulting in a 'randomization' (distortion) in the internal electric field intensity distribution that is manifested as a distortion in the far-field diffraction pattern.

A critical question in the analysis of 2-D diffraction patterns from composite microparticles is what is the minimum domain size that can be detected with this technique. Light scattering and resolution analysis of Fabry-Perot interferometers suggest that refractive index discontinuities (phase-separated domains) with a length scale $\lambda/20$ (25 – 30 nm), where λ is the laser wavelength, are required tc produce a measurable distortion in the diffraction pattern. If true, this "resolution" is comparable to radii of gyration for most large molecular weight polymers, thus providing a method of looking "within" a composite particle and probing material homogeneity on a molecular scale. Here we examine in some detail the question of domain size and number density of phase-separated sub-domains within a host particle.

We have considered the issue of domain size resolution in 2-dimensional optical diffraction of polymer-blend and polymer-composite microparticles both experimentally and theoretically. The question can be phrased in two parts: (1) What is the smallest phase-separated sub-volume (domain) of the particle that will manifest its presence as a distortion in the 2-dimensional diffraction pattern? (2) What is the sensitivity to number density or relative weight fraction? That is, if two polymers phase-separate in a microparticle, at what relative weight fraction (or number density of phase-separated inclusions) will one to be able observe phase separation (0.1, 1, 10% etc.) as a distortion in 2-D diffraction patterns? These questions are not trivial to address by first-principles electrodynamics, although a perturbative volume-current method has recently been employed to simulate distortion of 2-D diffraction patterns from binary particles.[33] These calculations suggest that, under the right experimental conditions, very low relative weight fractions ($< 1\%$) are required to observe distortion, with domain size resolution on the order of 20 – 40 nm depending on the relative difference in refractive index between host and guest material.

Experimentally, we find good agreement between experiment and theory with respect to domain size but somewhat poorer agreement in number density (relative weight fraction) requirements for diffraction distortion. We examined diffraction from polyethylene glycol host particles doped with varying weight fractions of ceramic nanoparticles (Al_2O_3, TiO_2) and 14-nm latex beads. The Al_2O_3 and TiO_2 particles have nominal sizes of 46 and 28 nm respectively (specified by manufacturer), but their refractive indices (1.57, and 2.1 respectively) are such that the two particles introduce approximately the same optical phase shift. Figure 4 shows a surface plot of 2-D diffraction data from a PEG host particle (7.5 μm diameter) doped with 46-nm Al_2O_3 particles at a 14:1 relative weight ratio. The intensity oscillations along the polar angles (left abscissa) are the signature of material inhomogeneity. Similar results were obtained for TiO_2. The dopant ceramic particles do indeed produce measurable distortion in the diffraction patterns (quantitatively described by Fourier transform of individual diffraction fringes), but only at relative weight fractions > 5%. Interestingly (see subsection 4.1.4), addition of 14-nm latex beads to PEG host particles did not result in distortion of the diffraction pattern at relative weight fractions up to 50%, but do significantly modify the refractive index.

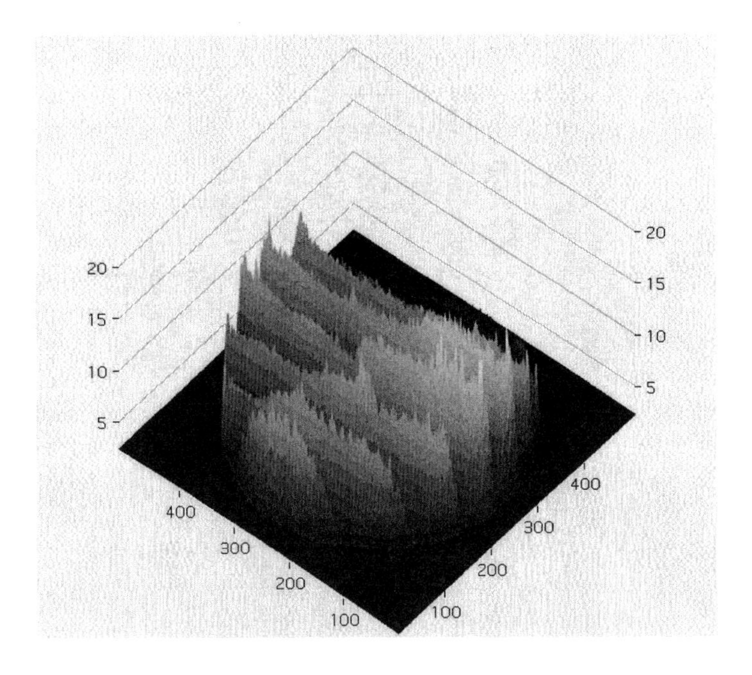

Figure 4. Surface plot representation of 2-D diffraction data from a PEG/Al_2O_3 (46 nm nom. diameter) composite particle in a 14:1 relative weight ratio. The x and y cooordinates are pixel numbers corresponding to azimuthal and polar scattering angles respectively. The host particle diameter is 7.5 μm. The intensity oscillations along each diffraction fringe signal material inhomogeneity.

Formation of homogeneous polymer-blend composites from bulk-immiscible co-dissolved components using droplet techniques has two requirements. First, solvent evaporation must occur on a relatively time scale compared to polymer translational diffusion. Second, the polymer mobility must be low enough so that, once the solvent has evaporated, the polymers cannot overcome the surface energy barrier and phase-separate. We have shown definitively the effects of droplet size and solvent evaporation, and the second requirement is almost always satisfied even for modest molecular weight polymers. In order to explore effects of polymer mobility in more detail, we looked at composite particles of PEG oligomers (MW 200, 400, 1000, and 3400) with medium molecular weight (14 k) atactic polyvinyl alcohol (PVA). This system allows us to systematically examine the phase separation behavior where one component (PEG) has substantially different viscosities (specified as 4.3, 7.3, and 90 cStokes at room temperature for PEG [200], PEG [400], and PEG [3400] respectively). All of the polymers examined in this study we probed at temperatures above the glass transition. The glass transition temperatures for PEG(400) and PEG(3400) are 205° and 232° K respectively.[1] - we did not find data for PEG(200) but expect it to be less than or approximately equal to that of PEG(400). So in the absence of deep evaporative cooling (which we have no evidence of) all the polymers studied are above their respective T_g.

We observed that the higher molecular weight PEG polymer-blend particles were homogeneous as determined from bright-field microscopy, optical diffraction and fluorescence imaging. Blend-particles prepared with the 200-molecular weight PEG were observed to form sphere-within-a-sphere particles with a PVA central core. On a time scale of about 10 minutes, the composite particle undergoes phase separation into an inhomogeneous particle as evidenced by fringe distortion. Interestingly, the structure in the 2D diffraction data for this system is much different than those observed for large phase-separated PVC/PS particles that presumably coalesce into sub-micron spheroidal domains.[34] Based on fluorescence imaging data, [35] the PEG[200]/PVA[14k] particle forms spherically symmetric (sphere-within-a-sphere) heterogeneous structures, which *should* also (but does not) produce well-defined diffraction fringes.[36,37,38]

Our interpretation of this observation is that diffusional motion of the PVA core in the PEG host particle, combined with rotational diffusion of the particle breaks the spherical symmetry and thereby introduces distortion in the diffraction pattern. This observation is entirely consistent with our model of polymer-composite formation where heterogeneous particles may be formed provided that the mobility of one of the polymers is low enough to overcome the surface energy barrier. Composite particles formed from the higher molecular weight PEG (>1000) form homogeneous blend particles with PVA. The time scale for transitions from a homogeneous to inhomogeneous particle provides an estimate (at least to an order of magnitude) of the diffusion coefficient of the light component *within* the particle. This is a number that can be connected directly to molecular dynamics simulations. By equating the average diffusion distance $r = (6Dt)^{1/2}$ to the particle radius (6 μm) and using a value of $t = 10$ minutes (600 s), we estimate a value of $D = 10^{-10}$ cm^2/s, which is consistent with recent molecular modeling results (see the following section). Composite particles formed from the higher molecular weight PEG (>1000) form homogeneous composite particles with PVA.

SUMMARY AND FUTURE DIRECTIONS

The combination of experimental evidence and computational modeling show conclusively that stable, homogeneously blended (bulk-immiscible) mixed-polymer composites can be formed in a single microparticle of variable size. To our knowledge, this represents a new method for suppressing phase-separation in polymer-blend systems without compatibilizers that allows formation of polymer composite micro- and nanoparticles with tunable properties such as dielectric constant. Conditions of rapid solvent evaporation (e.g. small (<10 μm) droplets or high vapor pressure solvents) and low polymer mobility must be satisfied in order to form homogeneous particles. While this work was obviously focused on polymeric systems, it should be pointed out that the technique is easily adaptable to making particles of small organic and inorganic (and hybrid composites) as well. A wide range of electronic, optical, physical and mechanical properties of single- and multi-component polymer nanoparticles remain to be explored.

Currently, we are investigating structure and dynamics of a number of polymer-polymer, and polymer-inorganic composite systems. The latter, for example, show an interesting time dependence in both size, morphology, and phase-separation behavior whose underlying microscopic mechanism is not yet completely clear. In other work, we are exploring the role of compatibilizers (short-chain copolymers) to increase the size range for producing homogeneous polymer-blend particles. Also, we are developing techniques for multi-color diffraction to extract information on ternary and higher-order composite particles. Some important issues relevant to commercialization remain to be resolved as well. Most importantly is the problem of particle throughput that is currently (optimistically) limited to a few milligrams per day. Other issues include hardware compatibility with various solvents that are commonly encountered in polymer solution work. Low-vapor pressure solvents such as methylene chloride are seriously problematic in acoustically driven droplet ejection devices such as ours. Effects such as cavitation, and clogging due to solvent evaporation will need to be confronted in order to expand the practical range of materials amenable with this technique.

ACKNOWLEDGEMENTS

This research was sponsored by the U.S. Department of Energy, Office of Basic Energy Sciences (Divisions of Chemical Sciences and Materials Science), under contract DE-AC05-96OR22464 with Oak Ridge National Laboratory, and Laboratory-Directed Research and Development Seed Money Fund managed by Lockheed Martin Energy Research Corporation. K. Fukui and J. V. Ford acknowledge support from the ORNL Postdoctoral Research Program. J. U. Otaigbe acknowledges support from the National Science Foundation under contract DMR9982077. We also acknowledge important contributions from Professor K. C. Ng (ORNL Faculty Research Participation Program) and K. Runge.

REFERENCES

[1] S. A. Jenekhe, X. J. Zhang, X. L. Chen, V. E. Choong, Y. L. Gao, and B. R. Hsieh, Chem. Mater. **9** 409-413 (1997).

[2] R. M. Tarkka, X. J. Zhang, and S. A, Jenekhe, J. Am. Chem. Soc. **118** 9438-9439 (1996).

[3] F. Croce, G. B. Appetecchi, L. Persi, and B. Scrosati, Nature **394**, 456-458 (1998).

[4] R. E. Schwerzel, K. B. Spahr, J. P. Kurmer, V. E. Wood, and J. A. Jenkins J. Phys. Chem. A **102** 5622-5626 (1998).

[5] M. Xanthos, Poly. Eng. Sci. 28, 1392 (1988).

[6] C. Koning, M. van Duin, C. Pagnoulle, and R. Jerome, Prog. Polym Sci. **23**, 707-757 (1998).

[7] D. R. Paul in *Polymer Blends*, Vol. II ed. D. R. Paul and S. Newman, Academic Press, New York, pp 35-62 (1978).

[8] L. A. Utracki, Polymer Alloys and Blends: Thermodynamics and Rheology, Oxford University Press, New York (1990).

[9] L. Sung, A. Karim, J. F. Douglas, and C. C. Han, Phys Rev. Lett **76**, 4368-4371 (1996).

[10] J. W. Yu, J. F. Douglas, E. K. Hobbie, S. Kim, and C. C. Han, Phys Rev. Lett. **78**, 2664-2667 (1997).

[11] A. H. Marcus, D. M. Hussey, N. A. Diachun, and M. D. Fayer, J. Chem. Phys. **103**, 8189-8200 (1996).

[12] S. A. Jenekhe, and X. L. Chen, Science **279**, 1903-1906 (1998); S. A. Jenekhe, and X. L. Chen *ibid.* **283** 372-375 (1999).

[13] F. S. Bates, and G. H. Fredickson, Phys. Today **52**, 32 – 38 (1999), and references cited therein.

[14] M. D. Barnes, C-Y. Kung, K. Fukui, B. G. Sumpter, D. W. Noid, and J. U. Otaigbe, *Optics Letters* **24**, 121-123 (1999).

[15] C-Y. Kung, M. D. Barnes, N. Lermer, W. B. Whitten, and J. M. Ramsey, *Analytical Chemistry* **70**, 658 - 661 (1998).

[16] C-Y. Kung, M. D. Barnes, N. Lermer, W. B. Whitten, and J. M. Ramsey, *Applied Optics* **38**, 1481 – 1487 (1999).

[17] M. D. Barnes, K. C. Ng, K. Fukui, B. G. Sumpter, and D. W. Noid, *Macromolecules* **32**, 7183-7189 (1999).

[18] We use the word "droplet" to denote liquid phase, and "particle" to denote a solid (dry) phase.

[19] R. Chang, and E. J. Davis, J. Coll. and Inter. Sci. **54**, 352 (1976).

[20] E. J. Davis, and A. K. Ray, J. Coll. and Inter. Sci. **75**, 566 (1980).

[21] A. K. Ray, A. Souyri, E. J. Davis, and T. M. Allen, Appl. Optics **30**, 3974 (1991).

[22] J. F. Widmann, C. L. Aardahl, and E. J. Davis Am. Lab. **28** 35 (1996).

[23] See also, the review by E. J. Davis, Aer. Sci. Tech. 26 212-254 (1997)

[24] P. Chylek, V. Ramaswamy, A. Ashkin, and J. M. Dziedzic, Appl. Optics **22**, 2302 (1983).

[25] H. C. Van de Hulst, and R. T. Wang, Appl. Optics **30**, 4755 (1991).

[26] G. Konig, K. Anders, A. Frohn, J. Aerosol Sci., **17**, 157 (1986).

[27] A. R. Glover, S. M. Skippon, and R. D. Boyle, Appl. Optics, **34**, 8409 (1995).

[28] J. F. Widmann, and E. J. Davis, Colloid and Polymer Science **274**, 525-531 (1996).

[29] T. Kaiser, S. Lange, and G. Schweiger Appl. Optics **33** 7789-7797 (1994).

[30] S. Holler, Y. Pan, R. K. Chang, J. R. Bottiger, S. C. Hill, and D. B. Hillis, Optics Lett. **23**, 1489 - 1491 (1998).

[31] J. F. Widmann, C. L. Aardahl, T. J. Johnson, and E. J. Davis, J. Coll. Int. Sci. **199**, 197 – 205 (1998).

[32] M. D. Barnes, N. Lermer, W. B. Whitten, J. M. Ramsey, Rev. Sci. Instrum. **68**, 2287 - 2291 (1997).

[33] J. V. Ford, B. G. Sumpter, D. W. Noid, M. D. Barnes, S. C. Hill, and D. B. Hillis, *J. Phys. Chem.* B **104**, 495-502 (2000).

[34] It should be noted that for heterogeneous particles, while the diffraction fringes are highly distorted, there remains some definite two-dimensional structure. This implies some uniformity and order of subdomains, however inversion of this type of data to extract such information is not trivial. This problem is currently under investigation. (See also Ref. 30).

[35] M. D. Barnes, K. C. Ng, K. P. McNamara, C-Y. Kung, J. M. Ramsey and S. C. Hill, *Cytometry* **36**, 169-175 (1999).

[36] J. A. Lock Appl. Optics **29** 3180-3187 (1990).

[37] A. K. Ray, B. Devakottai, A. Souyri, and J. L. Huckaby, Langmuir 7 525-531 (1991).

[38] R. L Hightower, C. B. Richardson, H-B. Lin, J. D. Eversole, and A. J. Campillo Optics Lett. **13** 946-948 (1988).

MOLECULAR SIMULATION AND MODELING OF THE STRUCTURE AND PROPERTIES OF POLYMER NANO-PARTICLES

Bobby G. Sumpter, Kazuhiko Fukui, Michael D. Barnes and Donald W. Noid
Chemical and Analytical Sciences Division
Oak Ridge National Laboratory
Oak Ridge, TN 37831-6197

INTRODUCTION

Recently we have developed an experimental technique for creating polymer particles of arbitrary composition and size.[1,2] In the experiment, instrumentation for generation and characterization of single molecules in solution of droplet streams with small (1-2 um) average diameter and monodispersity, was used.[3] This technique makes the initial volume of dilute solution sufficiently small so that the solvent evaporates on a very short time scale (few milliseconds). With a droplet generator specially designed and constructed in our laboratory, we have succeeded in generating and efficiently probing (i.e., reproducible delivery to a 1 μm target) 1-μm diameter water droplets at rates of up to 100 Hz. While the monodispersity is a function of initial droplet size, 1% relative standard deviation in particle size is routinely achievable with our on-demand droplet generator design. Using this technique, we generated polymer particles of arbitrary composition and size. Figure 1 shows Fraunhofer diffraction data from three different polyethylene oxide (PEO) microparticles, along with fits to the data using Mie theory.[15)] For micro- and nano-scale generated polymer particles, the refractive index is consistent with bulk (nominal) values, and the level of agreement with Mie theory indicates that the particles are nearly perfect spheres.[4] Figure 2(a) shows a schematic image of the generation of these polymer particles, and an optical image of typical particles formed from the experiment a droplet generator is shown in Fig. 2b.

Polymer particles in nano- and micrometer size range provide many unique properties due to size reduction to the point where critical length scales of physical phenomena become comparable to or larger than the size of the structure. Applications of such particles take advantage of high surface area and confinement effects, which leads to nano-structures with different properties than conventional materials. Clearly such changes offer extraordinary potential for development of new materials in the form of bulk composites and blends which can be used for coatings, opto-electronic components, magnetic media, ceramics and special metals, micro- or nano-manufacturing, and bioengineering.[5] In order to develop an understanding of properties influenced by the size scale of the materials, we have begun an extensive set of computational experiments.

A computational algorithm for generating and modeling polymer particles was developed

Computational Studies, Nanotechnology, and Solution Thermodynamics of Polymer Systems
Edited by Dadmun *et al.*, Kluwer Academic/Plenum Publishers, New York, 2000

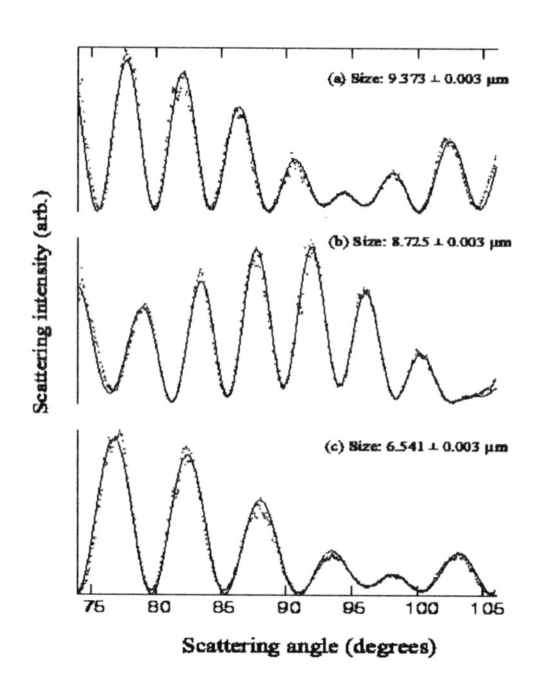

Figure 1 Experimental and calculated 1-dimensional Franhofer diffraction patterns from electrodynamically levitated polyethylene oxide (PEO) particles produced in situ with an on-demand droplet generator. The PEO weight fractions in water were 3, 2.5 and 1% for a, b, and c respectively. The refractive index (1.461 ± 0.001) determined from the data analysis is in good agreement with the refractive index of bulk (10K molecular weight) PEO.

for constructing particles that are as similar as possible to the experimentally created polymer particles. We have examined a variety of PE nano-scale particles, allowing the systematic study of size-dependent physical properties of these particles.[6-9] The models have been well tested and shown to provide realistic representation of the structure and vibrational spectroscopy of a number of polymer systems.[10]

MODEL DEVELOPMENT AND RESULTS

The molecular dynamics method is well known and has been reviewed in several papers.[11,12] Basically one needs to solve Hamilton's equations, or any other formulation of the classical equations of motion, starting from some initial positions and momenta of all of the atoms in the system and propagating the solution in a series of time steps. In our MD simulations, the integrations of the equations of motion are carried out in Cartesian coordinates, thus giving an exact definition of the kinetic energy and coupling. The classical equations of motion are formulated using our geometric statement function approach,[13] which significantly reduces the number of mathematical operations required. These coupled equations are solved using novel symplectic integrators developed in our laboratory which conserve the volume of phase space and robustly allow integration for virtually any time scale.[14]

Molecular Hamiltonian and Potential Energy Functions

For simplicity, the united atom PE model was used in which the CH_2 and CH_3 groups are treated as a single particle (bead) of mass m = 14.5 amu. By neglecting the internal structure of those groups, we reduce the number of equations of motion for the system and save a significant amount of computational time. The MD simulations compute the momenta and coordinate of

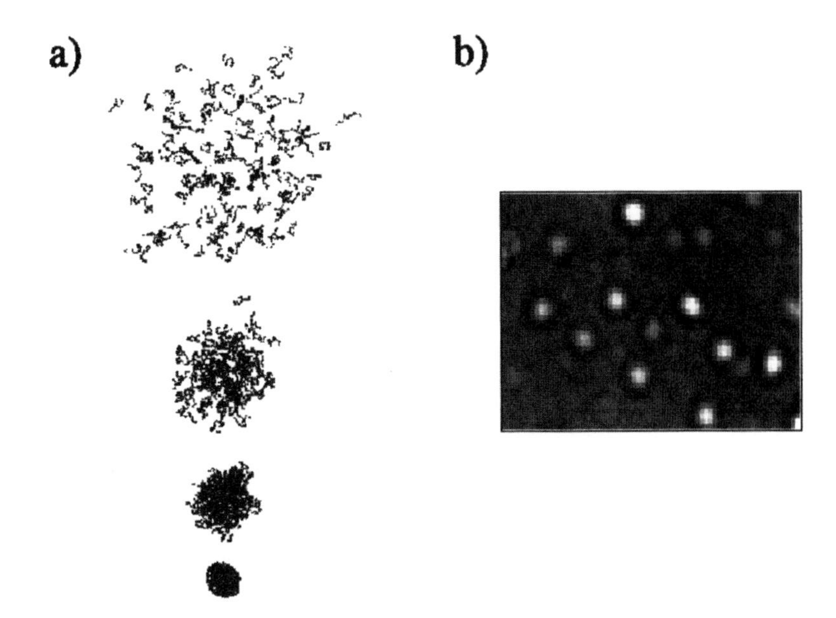

Figure 2 a) A molecular dynamics simulation was used to create a schematic image of polymer nano- and macro-scale particles generated from the submicron liquid droplets experiment. The solvent molecules (typically H_20 and THF in the experiment, not considered in the simulation) are streaming away as the polymer molecules are left behind collapsing into a nano or micrometer particle. Within the experimental time scale, the solvent evaporation was completed before the particles collected on a substrate. b) Bright field image (Inverted grayscale) of 2 ìm diameter of PE particles.

each atom in the system as a function of time by integrating Hamilton's equations of motion using the molecular Hamiltonian written in Cartesian coordinates. A classical molecular Hamiltonian for a polymer system can be written in terms of the kinetic and potential energy of the system

$$H = \sum_{i=1}^{N} \frac{\boldsymbol{p}_i^2}{2m} + \sum V_{2b}(r) + \sum V_{3b}(\theta) + \sum V_{4b}(\tau) + \sum_{i=1}^{N-3} \sum_{j \geq i+3}^{N} V_{Nb}(r_{ij}) \quad , \tag{1}$$

where N is a total number of atoms and $\boldsymbol{p}_i$ is the Cartesian momentum of the ith atom. The r, θ and τ are the internal coordinates for the interatomic distance, the bending angle between three consecutive atoms, and the torsional angle between four consecutive atoms, respectively. For the two-body bonded lengths and the three-body bending angles, simple harmonic potential functions are used,

$$V_{2b}(r_{i,i+1}) = \frac{1}{2} k_r \, (r_{i,i+1} - r_0)^2 \tag{2}$$

where k_r and k_θ are the force constants parameters. The r_0 and θ_0 are the equilibrium value of the bond length and the angle formed by the three atoms of interest in a particular bond. The

$$V_{3b}(\theta_{i,i+1,i+2}) = \frac{1}{2} k_\theta (\theta_{i,i+1,i+2} - \theta_0)^2 \tag{3}$$

torsional potential for the four body angles, τ, is given by

$$V_{4b}(\tau_{i,i+1,i+2,i+3}) = 8.77 + \alpha \cos\tau_{i,i+1,i+2,i+3} + \beta \cos^3\tau_{i,i+1,i+2,i+3} \tag{4}$$

The four-body torsional potential term was developed by Boyd and gives a cis barrier of 16.7 KJ/mol and a gauche-trans energy difference of 2.5 KJ/mol.[15] The values of α and β are potential parameters for the barrier to intra-molecular rotational isomerization. The last term of the potential energy functions in Eq. (1) is the nonbonded two-body interaction (the distance for cutting off the interaction is set at 10 Å). The function for the interaction between any two atoms not directly bonded together (1-4 types and larger, j i+3) is:

$$V_{\mathrm{Nb}}(r_{i,j}) = 4\epsilon \left[\left(\frac{\sigma}{r_{i,j}} \right)^{12} - \left(\frac{\sigma}{r_{i,j}} \right)^{6} \right] \tag{5}$$

where ϵ and σ represent the Lennard-Jones parameters. The parameters for the potential energy functions of Eq.(2)-(5) are shown in Table 1.

TABLE 1. Potential parameters

Stretch	r_0	k_r
C-C	1.53	2651 kJ/mol
Bending	θ_0	k_θ
C-C-C	113°	130.1 kJ/mol
Torsion	α	β
C-C-C-C	-18.41 kJ/mol	26.78 kJ/mol
Nonbonded	ε	σ
CH_2 CH_2	0.494 kJ/mol	3.98

Modeling of Polymer Particles

We have developed an efficient method to obtain a desired particle size to model production of polymer particles using MD simulations.[7] The procedure starts by preparing a set of randomly coiled chains with a chain length of 100 beads by propagating a classical trajectory with a perfectly planar all-trans zig-zag initial conformation and randomly chosen momentum with a temperature of 300 K. The trajectory is terminated at 200 ps, and position and momenta of the chain are saved. By repeating this process, a desired set of randomly coiled chains is prepared. From the set, six chains are selected and placed along the Cartesian Axis. To create a collision at the Cartesian origin, these chains are propelled with an appropriate amount of momentum, the resulting particle consisting of the six chains is annealed to a desired temperature and rotated through a randomly chosen set of angles in three-dimensional space to create a homogeneous particle. This process is continued until the desired size is achieved for the study. Figure 3 illustrates the process for generating initial conditions of different sized PE particles.

We also generated the PE particles with various chain lengths (50 and 200 beads) based on the initial configurations of the generated amorphous PE particles with a chain length of 100 beads. To do this, we simply expand the space of the particle with a chain length of 100 beads by multiplying the initial configuration by two in Cartesian coordinates. Then, a new atom is inserted between those two atoms. Thus, the total number of atoms is doubled and the new particle consists of chains with a length of 200 beads. For the particle with a chain length of 50 beads, the two, three and four-body bonded interaction are simply turned off every 50 beads. Finally, classical trajectories are propagated with randomly chosen momenta until the density reaches 0.7 g/cm^3 and annealed to a desired temperature. It is noted that the density for the particle is low (0.05 g/cm^3) and the temperature is very high (1000 K) at the start of those trajectories, since the space of the original particle is expanded. Using this scheme, we can efficiently create particles with various chain lengths without propelling sets of six chains.

After a desired size PE particle is obtained, classical trajectories are propagated for 50 ps at the above bulk melting point, and then annealed by scaling the Cartesian momenta with a constant scaling factor until the temperature reaches 10 K to find a steady state of the amorphous PE particles. To obtain average values of properties of the particles at a fixed temperature and examine dependence of the conformations on temperature, we have used Nosé-Hoover chain (NHC) constant temperature molecular dynamics.[16] The initial configurations of the steady state of the amorphous PE particle are used at the start of the NHC simulations; the initial values of the Cartesian momenta are given random orientation in phase space with magnitudes chosen so that the total kinetic energy is the equipartition theorem expectation value.[17] The temperature of the particle T is calculated from

$$\frac{3}{2} N k_B \langle T \rangle = \left\langle \sum_{i=1}^{N} \frac{p_i^2}{2m} \right\rangle \tag{6}$$

where k_B is the Boltzmann constant and N is the total number of atoms. After we propagate NHC trajectories 10 ps to equilibrate the system at a desired temperature, we begin sampling the molecular positions and momenta at a uniform interval (1.0 ps) until the simulation time reaches 100 ps. Figure 3 d) shows polyethylene particles (60,000 atoms) with chain length of 100 beads at a temperature of 10 K.

Modeling of Bulk Polymer Systems

Modeling of a polyethylene bulk system is achieved by placing 32 CH_2 chains with 100 C and 202 H atoms in a cubic box under three-dimensional periodic boundary conditions. The initial configuration of the system was optimized using molecular mechanics calculation. With the initial conformation, MD simulations are propagated by applying external force to the box every 2 fs in order to pack the chains into the box with the desired density. The simulation is terminated when the density reaches 0.85 g/cm^3 (bulk amorphous PE density). For the simulations of the bulk system, we considered all atoms of PE polymer in order to compare the computed structural conformations (*e.g.* radial distribution) with X-ray and neutron scattering experiments. The radial distribution of the simulated bulk PE was in good agreement with the experimental data. By comparing the bulk system with the nano-scale particles, we can study the conformational change of the particles due to the size reduction and the shape.

Nano-Particle Structural Characteristics

We modeled polymer nano-particles which should closely correspond to the ones that can be experimentally created from a droplet generator that efficiently generates nearly perfect spherical sub-micron particles of arbitrary composition. Using the molecular dynamics technique, PE particles with chain lengths of 100 beads generated with up to 60,000 atoms have almost a spherical shape (a symmetry parameter -0.1 at a temperature of 10 K) as shown in Fig. 3b, in good agreement with our experimental results. To interpret properties of the polymer fine particles differing from their bulk solid phase, we first counted the surface atoms using a three-dimensional grid method in Cartesian coordinates and the ratio of the surface atoms to the total number of atoms are obtained (the diameter is the average value of distance from a center of mass to the surface atoms). Since the smaller sized particles have more surface atoms than the larger ones, a decrease of the diameter increases the ratio as shown in Table 2. The large ratio of surface atoms to the total number of atoms provides reduction of the nonbonded interactions for the surface layer; hence the cohesive energy is dramatically dependent on the size. In addition, the ratio of surface chain ends to total number of chain ends for the particles is much larger than that of the bulk system, leading to enrichment of chain ends at surface. This observation is consistent with analysis of thin films.[18-20] With regard to an effect of the side-atoms, the increase in the side-atoms corresponds to a decrease in the ratio of surface atoms and therefore represents an increase of cohesive energy of the system.

The surface area and volume are calculated using the contact-reentrant surface method[21] with a probe radii, $R_p = 1.4$ Å. The large proportional surface area defined by S_{ratio} = (surface area)/(volume) leads to large surface free energy, which is described by per unit of surface area (J/nm^2). The S_{ratio} of the particles is large compared with that of the bulk so that the surface area and surface free energy are large. The surface of the particles is also characterized by the fractal dimension which describes a degree of irregularity of a surface.[22] The values of D are smaller for the particles than the value of the bulk ($D = 2.72$). This indicates that the surface is irregular and has many cavities which may introduce unique (catalytic or interpenetrating) properties of polymer fine particles. This predicts that nano-scale polymer particles are loosely packed and can show dynamical flexibility (e.g. Compressive modulus of the particles is much smaller than that of the bulk system). The free volume (cavities) and molecular packing can be important in

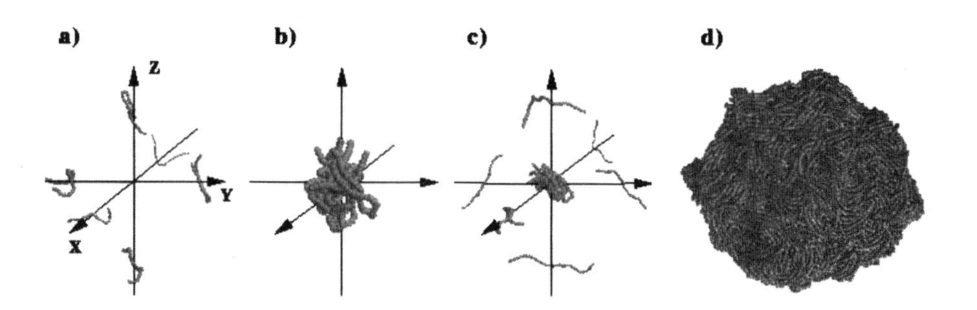

Figure 3 Creation of polyethylene droplets for initial conditions: a) randomly coiled 6 chains with chain length of 100 beads, b) a particle for 600 atoms with those 6 chains, and c) the particle and a new set of 6 chains. Before the set is slung, the particle was randomly rotated. d) 60000 atom of PE particle with a chain length of 100 beads

a diffusion rate of a small molecule trapped in the particles.[23] The structural characteristics of the PE particles up to 60,000 atoms with a chain length of 100 beads are shown in Table 2.

TABLE 2. Characteristics of polymer nano-particles

No. of atoms	Diameter (nm)	Ratio of chain ends	Ratio of surface atoms	Area [a] (nm^2)	Volume [a] (nm^3)	S_{ration} (nm^{-1})	Fractal Dim. (nm^{-1}) [b]
3000	4.7	95%	72 %	152	67	2.27	2.14
6000	5.6	87%	70 %	257	138	1.86	2.15
9000	6.4	85%	69 %	346	209	1.65	2.16
12000	7.0	83%	67 %	443	279	1.55	2.16
18000	8.1	79%	65 %	610	419	1.46	2.19
24000	9.0	76%	64 %	814	564	1.44	2.21
30000	9.7	71%	62 %	955	706	1.35	2.21
60000	12.4	69%	60 %	1879	1424	1.32	2.26

[a] The probe radius is set at 1.4 Å.
[b] The probe radius is in the range from 2.0 to 3.5 Å.

Figure 4 shows radial distribution functions for the PE particles of 12,000, 30,000 and 60,000 atoms with a chain length of 100 beads at a temperature of 100 K. For the amorphous PE particles, the peak positions of the radial distributions are insensitive to the size in the diameter range 12.5 nm. By comparing the peak positions of the radial distributions of the particles with the bulk system, it is clear that the peak around 3.15 Å corresponding to *gauche* configuration is very small for the particles. The reduction of *gauche* configuration in the radial distributions is believed to be due to alignments of the chains on the surface. In our previous study, we monitored the averaged end-to-end distances of the surface- and inner-chains for the particle of 12,000 atoms with a chain length of 100 beads. The average end-to-end distance of the surface chains is longer than that of the inner chains, and the inner chains have more gauche configurations than the surface.[6]

Several simulations have been applied to study the morphology of single or multiple chains with different chain lengths.[24,25] Since the surface chains of the PE nano-particles tend to straighten and aline at temperatures below the melting point, the preferential morphology for the small PE particle with a long chain length is a rod-like shape. This mechanism was also observed by Liu and Muthukumar in the simulations of polymer crystallization.[25] This stretching of the chains leads to reduction of the cohesive energy and an increase in volume. Studies on the effect of a chain length show that the particles with the shortest chain length (50 beads) have the most spherical shape.

Thermal Properties of PE Nano-particles

Thermal analysis provides a great deal of practical and important information about the

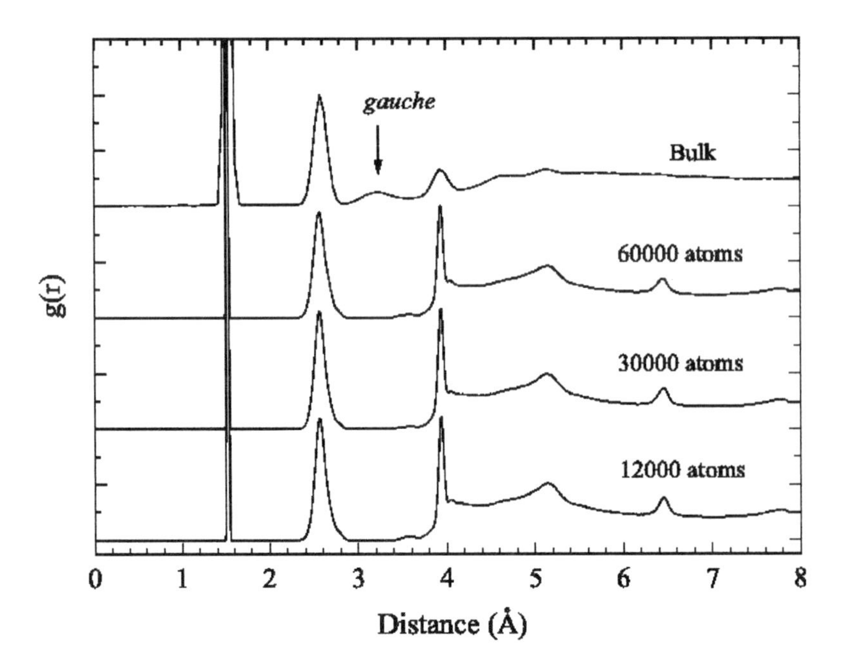

Figure 4 Radial distribution functions for several different polymer particles as compared to that of the bulk system.

molecular and materials world relating to equations of state, critical points and the other thermodynamics quantities. For our study of thermodynamic properties of nano-scale polymer particles, we have calculated temperature, volume, and total energy in the process of annealing the system by scaling the momenta at a constant rate. To determine the appropriate annealing schedule[7], we have examined the melting point and glass transition temperature for the particle of 12,000 atoms with a chain length of 100 beads for different rates in the rage from 1.7 K/ps to 29.5 K/ps. The transition temperatures were found to be rather insensitive (within the error of 5 K) to annealing rates slower than 6.0 K/ps. In all subsequent simulations, we set the annealing rate at approximately 2.5 K/ps so that the particle is thermally equilibrated for each sampling point. Computing the straight lines of total energy of the system or molecular volume *vs.* temperature with a least square fit, we take the points where those extrapolated straight lines meet, as the melting point, T_m , and the glass transition temperature, T_g. Table 3 summarizes thermal properties of the PE particles with respect to size and various chain lengths.

Figure 5 shows dependence of melting point and glass transition temperature on the diameter of the particles. The dramatic reduction of the melting point for the fine polymer particles is a example of surface effects and shows the importance of size. Since the large ratio of surface atoms to the total number leads to a significant reduction of the non-bonded interactions, the melting point decreases with decrease of the total number of atoms. Figure 6 shows the effect of chain length on transition temperatures. A strong dependence of the melting point and the glass transition temperature on chain length is attributed to molecular weight and non-bonded energy of each chain.

TABLE 3. Thermal properties of polyethylene nano-particles

No. of atoms	Cohesive energy[a)] (Kcal/mol)	T_m (K)	T_g (K)	C_p (cal/mol K) $T \leq T_m$	$T \geq T_m$
		Chain length of 100 beads			
3000	3010	218	111	6.37	7.57
6000	6510	234	134	6.50	7.70
9000	10170	242	131	6.44	7.22
12000	13900	242	157	6.47	7.24
18000	23900	249	155	6.44	7.66
24000	32000	254	154	6.45	7.95
30000	41170	258	152	6.40	7.53
60000	82950	266	161	6.65	7.80
		Chain length of 50 beads			
6000	7510	186	73	6.20	7.23
12000	15300	218	110	6.37	7.74
18000	25300	220	117	6.37	7.96
24000	34400	232	125	6.53	7.83
		Chain length of 200 beads			
6000	5738	285	152	6.37	8.02
12000	10020	317	162	6.76	7.73
18000	18800	332	168	6.81	8.32
24000	25200	345	172	7.03	8.22
		Chain length of 400 beads			
6000	4059	304	165	6.22	7.57
12000	9580	323	166	6.42	8.00
		Chain length of 1000 beads			
6000	3112	331	169	6.22	8.17
12000	6186	349	171	6.96	8.35
		Bulk[b)]			
	9073	414	195	5.19	7.67

[a)]The values are calculated from NHC simulations at 10 K.
[b)]Cubic boundary conditions were used with a box length of 4.6 nm. The values of C_p are from Ref. 30 at 300 K and 400 K.

Mechanical properties

The compressive modulus, which is quantified by the relation between stress and strain, are investigated for the PE particles by applying an external force in the MD simulations. Figure 7 is a plot of the stress as a function of strain for a PE particle consisting of 12000 beads with a chain length of 100 beads. The slope calculated for the proportional range gives a modulus for compression of the polymer nano-particles. It is known that values of the tensile modulus of bulk polyethylene are between 210 and 340 GPa.[26] In general, the compressive modulus is higher than the tensile modulus.[27,28] In addition, the bulk and tensile strength or yield point are usually much smaller than the modulus for a thermoplastic such as polyethylene. In the MD study of the nano-particles, we have observed a compressive modulus that is orders of magnitude smaller than the

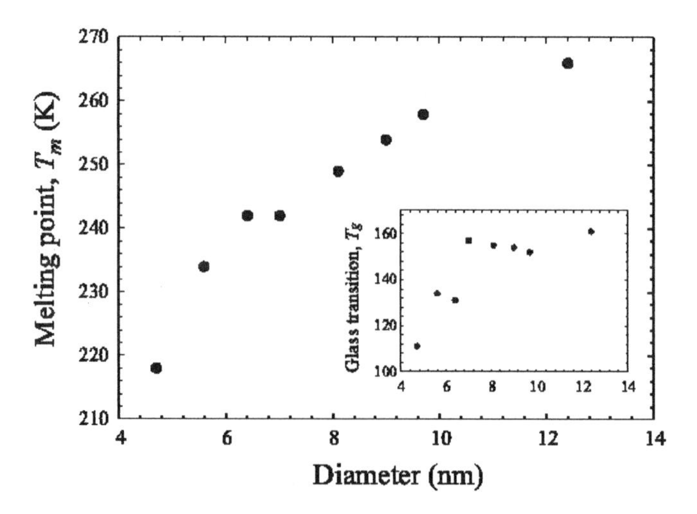

Figure 5 Dependence of the melting point and glass transition temperature (insert) on diameter for PE particles with a chain length of 100 beads.

bulk values and the yield point is much larger than the modulus. The stress-strain curve actually looks more like a curve for an elastomer. However, the initial deformation caused by the compression (that which gives the modulus) is not reversible (region A in Figure 7). What occurs during this phase is the deformation of a spherical particle to an oblate top. This structure is stable but it lies at a slightly higher energy than that for a spherical particle. Thus, the modulus for compression in this region is actually more a measure of the force required to deform the spherical polymer particle into an oblate top (pancake-like structure[1,9]). Further deformation tends to be reversible up to the point of rupture (region B in Figure 7). This deformation is actually more closely related to the bulk modulus since the stress is due to the cohesive energy and not a microstructure. This leads to a yield point that is significantly larger than the modulus. Table 4 shows the dependence of the mechanical properties of PE particles on size.

TABLE 4. Mechanical properties of PE nano-particles[a)]

No. of atoms	Compressive modulus (A) (MPa)	Compressive strength[b)] (MPa)	Compressive modulus (B) (GPa)
3000	68	288	46
6000	86	324	63
9000	88	327	67
12000	111	400	85

[a)]The modulus are calculated from the region of non-reversible and reversible deformation.(See Fig. 7, label A and B [b)]The strength are obtained at the point, M.

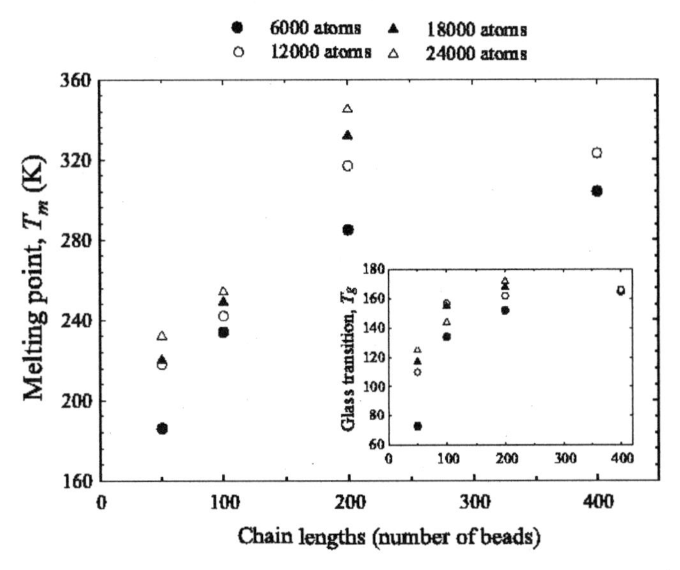

Figure 6 Dependence of the melting point and glass transition temperature (insert) on the chain length.

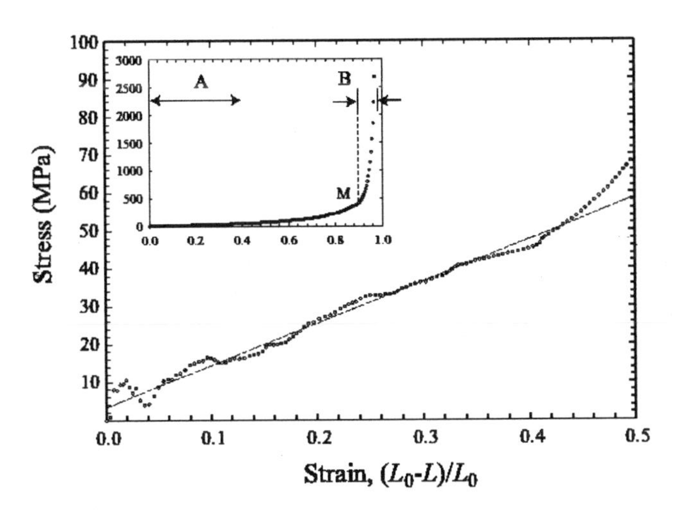

Figure 7. Stress-strain behavior for the PE particle of 12000 atoms with a chain length of 100 beads. The yield point is indicated at point M.. The region of non-reversible deformation caused by the compression is labeled A: the region of reversible deformation is labeled B. The slopes calculated in the strain rage from 0 to 0.4.(region A) and from 0.86 to 0.97 (region B) are 111 MPa and 85 GPa, respectively.

CONCLUSIONS

In this study of polymer nano-particles using molecular dynamics simulations we have analyzed the thermal and mechanical properties of the particles generated with up to 60,000 Two groups of atoms were singled out for analysis: those in the central and adjacent fluid slices at the beginning of the simulation. The fraction of kinetic energy in the z direction was followed in each group of atoms. Typical results are shown in Fig. 2. The fraction of z kinetic energy steadily decreases in the central region atoms, averaging about the thermal equilibrium value, 1/3, within less than 5 ps. The adjacent region atoms suddenly gain z kinetic energy within about the first 0.1 ps and then a decreasing trend similar to that for the central region atoms. No coherent transfer of z kinetic energy is observed. In short, the motion very rapidly thermalizes, which is consistent with the results of visualization. In several simulations, positions of every fluid atom were saved every 0.05 ps. The movies showed that within 2 ps, the atoms in the initial plug essentially interspersed themselves within the adjacent fluid volume and then randomly dispersed. This behavior was observed at each nanotube radius. atoms. It is found that surface effects provide interesting properties that are different from those of the bulk polymer system. In particular, the melting point and glass transition temperature were found to be dramatically dependent on the size of the polymer particles. This study also demonstrated that the nano-scale PE particles have dynamical flexibility and behave like an elastomer. The result is quantified by the fractal dimension and compressive modulus. We are currently extending this model and methodology to larger size particles and other types of polymer systems to study the interfacial tension between incompatible polymers, shear flow effects, and thermal properties of blended polymer particles.[29] The molecular dynamics simulations used here should provide useful insights to explain and predict the properties and behavior of ultra fine polymer particles to be used in future new materials and devices.

Acknowledgments

Research sponsored by the Division of materials Sciences, Office of Basic Energy Sciences, US Department of Energy under Contract DE-AC05-96OR22464 with Lockheed Martin Energy Research. K.F. is supported by the Postdoctoral Research Associates Program administered jointly by Oak Ridge National Laboratory and the Oak Ridge Institute for Science and Education.

References

(1) C-Y. Kung, M. D. Barnes, B. G. Sumpter, D. W. Noid and J. U. Otaigbe, Polymer Preprints ACS, Div. Polym. Chem. **(1998)** 39, 610.
(2) M. D. Barnes, C.-Y. Kung, N. Lermer, K. Fukui, B. G. Sumpter and D. W. Noid, Opt. Lett. **(1999)** 24, 121
(3) C-Y. Kung, M. D. Barnes, N. Lermer, W. B. Whitten and J. M. Ramsey, Anal. Chem. (**1998**) 70, 658.
(4) M. D. Barnes, N. Lermer, W. B. Whitten and J. M. Ramsey, Review of Scientific Instruments **(1997)** 68, 2287.
(5) C. Hayashi, R. Uyeda and A. Tasaki, *Ultra-Fine Particles Technology* (Noyes, New Jersey, **1997**).
(6) Fukui, K., Sumpter, B. G., Barnes, M. D., Noid, D. W. and Otaigbe, J. U., Macromol. Theory Simul. **(1999)** 8, 38.
(7) Fukui, K., Sumpter, B. G., Barnes, M. D. and Noid D. W., Comput. Theor. Polym. Sci. **(1999)** 9, 245.
(8) Fukui, K., Sumpter, B. G., Barnes, M. D. and Noid, D. W. Polym. J. **(1999)** 8, 664.
(9) Fukui, K., Sumpter, B. G., Barnes, M. D. and Noid, D. W., Chem. Phys. **(1999)** 244, 339.
(10) B. G. Sumpter, D.W. Noid and B. Wunderlich, J. Chem. Phys. **(1990)** 93, 6875.

(11) M. L. Klein, Annu. Rev. Phys. Chem. (**1985**) 36, 525.
(12) W. G. Hoover, Annu. Rev. Phys. Chem. (**1983**) 34, 103.
(13) D.W. Noid, B.G. Sumpter, B. Wunderlich and G.A. Pfeffer, J. Comp. Chem. (**1990**) 11, 23.
(14) S. K. Gray, D. W. Noid and B. G. Sumpter, J. Chem. Phys. (**1994**) 101, 4062.
(15) R. H. Boyd and S.M. Breitling, Macromolecules (**1974**) 7, 855.
(16) G. J. Martyna, M. L. Klein and M. Tuckerman, J. Chem. Phys. (**1992**) 97, 2635.
(17) K. Fukui, J. I. Cline, J. H. Frederick, J. Chem. Phys. (**1997**) 107, 4551.
(18) Dorulker, P.; Mattice, W. L., *Macromolecules* **1998**, 31, 1418.
(19) Kumar, S. K.; Vacatello, M.; Yoon, D.Y., *J. Chem. Phys.* **1988**, 89, 5206.
(20) Tanaka, K.; Takahara, A.; Kajiyama, T., *Macromolecules* **1997**, 30, 6626.
(21) M. L. Connolly, J. Am. Chem. Soc. (**1983**) 107, 1118.
(22) J. Doucet, J. Weber, *Computer-Aided Molecular Design: Theory and Applications* (Academic Press Inc., San Diego, **1996**).
(23) P.V. Krishna Pant, R. H. Boyd, Macromolecules (**1993**) 26, 679.
(24) G. Tanaka and W. L. Mattice, Macromolecules (**1995**) 28, 1049.
(25) C. Liu and M. Muthukumar, J. Chem. Phys. (**1995**) 103, 9053.
(26) Shoemaker, J., Horn, T., Haaland, P., Pachter, R. and Adams, W. W., Polymer, (**1992**) 33, 3351.
(27) Nielen, L. E., *Mechanical Properties of Polymers and Composites*, (Marcel Dekker, Inc., New York, **1974**).
(28) Porte, D., *Group Interaction Modelling of Polymer Properties* (Marcel Dekker, Inc., New York, **1995**).
(29) M. D. Barnes, K. C. Ng, K. Fukui, B. G. Sumpter, and D. W. Noid Macromolecules (**1999**) **32**, 7183.

THEORY OF THE PRODUCTION AND PROPERTIES OF POLYMER NANOPARTICLES: QUANTUM DROPS

Keith Runge, Kazuhiko Fukui, M. Alfred Akerman, M. D. Barnes, Bobby G. Sumpter, D. W. Noid

Chemical and Analytical Sciences Division, Oak Ridge National Laboratory, Oak Ridge, TN 37831-6197

INTRODUCTION

A recent development in the production of polymer particles has created a revolutionary new technology for the production of submicron polymer particles from solution.[1] In this experiment, generation and characterization of droplet streams with small (≤5 μm) average diameters have been used to create nano-polymer particles. This technique makes the initial volume of a dilute solution of any polymer material sufficiently small so that the solvent evaporation occurs on a very short time scale leaving behind a polymer particle. For micro and nano-scale generated polymer particles, the refractive index obtained from the data analysis is consistent with bulk (nominal) values and the level of agreement with Mie theory indicates that the particles are nearly perfect spheres.

We have previously presented results of calculations showing that polymer nanoparticles with excess electrons exhibit discrete electronic structure and chemical potential in close analog with semi-conductor quantum dots.[2,3] The dynamics of the formation of polymer nanoparticles can be simulated by the use of molecular dynamics and the morphology of these particles may be predicted. The production method that is used for the creation of these polymer particles can also be used to mix polymer components into a nanoparticle when otherwise they are immiscible in the bulk. Quantum drops, unlike the semiconductor quantum dots, can be generated on demand and obtained in the gas phase. In the gas phase, these new polymer nanoparticles have the capacity to be used for catalytic purposes which may involve the delivery of electrons with chosen chemical potential. Finally, quantum drops have unusual properties in magnetic and electric fields, which make them suitable for use in applications ranging from catalysis to quantum computation.

Computational Studies, Nanotechnology, and Solution Thermodynamics of Polymer Systems
Edited by Dadmun *et al.*, Kluwer Academic/Plenum Publishers, New York, 2000

SIMULATION OF POLYMER NANOPARTICLE FORMATION

A computational chemistry project has been initiated to parallel the new experimental technique recently developed by us at ORNL for creating very fine polymer particles of arbitrary composition and size. We have used molecular dynamics (MD) simulations of nano-scale polymer particles to gain insight into the properties and behavior of ultra fine polymer powders. A computational procedure to model polymer particles has been developed to generate initial particles that are as similar as possible to the experimentally created polymer particles. Using an efficient computational method to create initial conditions, we have modeled the nano-sized particles generated with up to 300,000 atoms under solvent free and vacuum conditions. Figure 1 illustrates the formation of stable polymer nano-particle from the collapsing chains when the solvent evaporates. The solvent molecules are not shown as they escape from the liquid drop so that the condensation of polymer chains can be visualized. We have simulated a variety of different polymers with different chain lengths and functionalities to investigate effects of structure and various statistical properties. Thermodynamic analyses have been performed to obtain melting point, glass transition temperature, and heat capacity for the particles by calculating molecular volume and total energy as a function of temperature and particle size. These analyses show that there is a surface effect, dramatically dependent on the size of the particles, and leading to such effects as greatly increased reactivity and modified spectra. Our simulations also predict an interesting reduction of the melting point and glass-transition temperature in comparison with the bulk system.

CALCULATION OF THE ELECTRONIC PROPERTIES OF QUANTUM DROPS

The current and emerging technologies for the production of these polymer nanoparticles generally result in the quantum drop having a residual static electric charge. The polarity of the residual charge may be either positive or negative, however, for the purposes of this contribution the residual charge will be taken to consist of an excess of electrons confined within the polymer nanoparticle. The computation of the electronic properties of quantum drops involves modeling a large number of atoms and electrons with no periodic behavior in any direction. To solve this problem we choose to implement a semi-classical calculation of the electronic states of polymer quantum drops, as has recently been applied to semiconductor quantum dots[4-9]. We examine a model Hamiltonian that includes the effect of electric and magnetic field, but which neglects the role of the electron spin.

It is clear from classical electrodynamics that the excess electrons would find their equilibrium configuration at the surface of the polymer quantum drop. For any deviation in the radial direction away from the surface of the drop, the electron would experience a restoring force that could be represented in the form of an image charge located a similar distance from the surface as the electron in the opposite radial direction. This classical electrodynamic model would provide the forces under which a single electron on the surface of a quantum drop would move. For the situation in which there is more than one electron, we need to take into account the electron-electron repulsion as well as the restoring force. We have developed a Hamiltonian that allows us to treat both the mutual repulsion of the electrons and, in an approximate way, the confinement of the electrons near the surface of the polymer quantum drop. The semi-classical energy spectrum of the system can be generated by a Poincaré surface-of-section technique.[10-13] Inclusion of electric and magnetic fields in the model Hamiltonian gives a first view into the unique electronic properties of quantum drops.

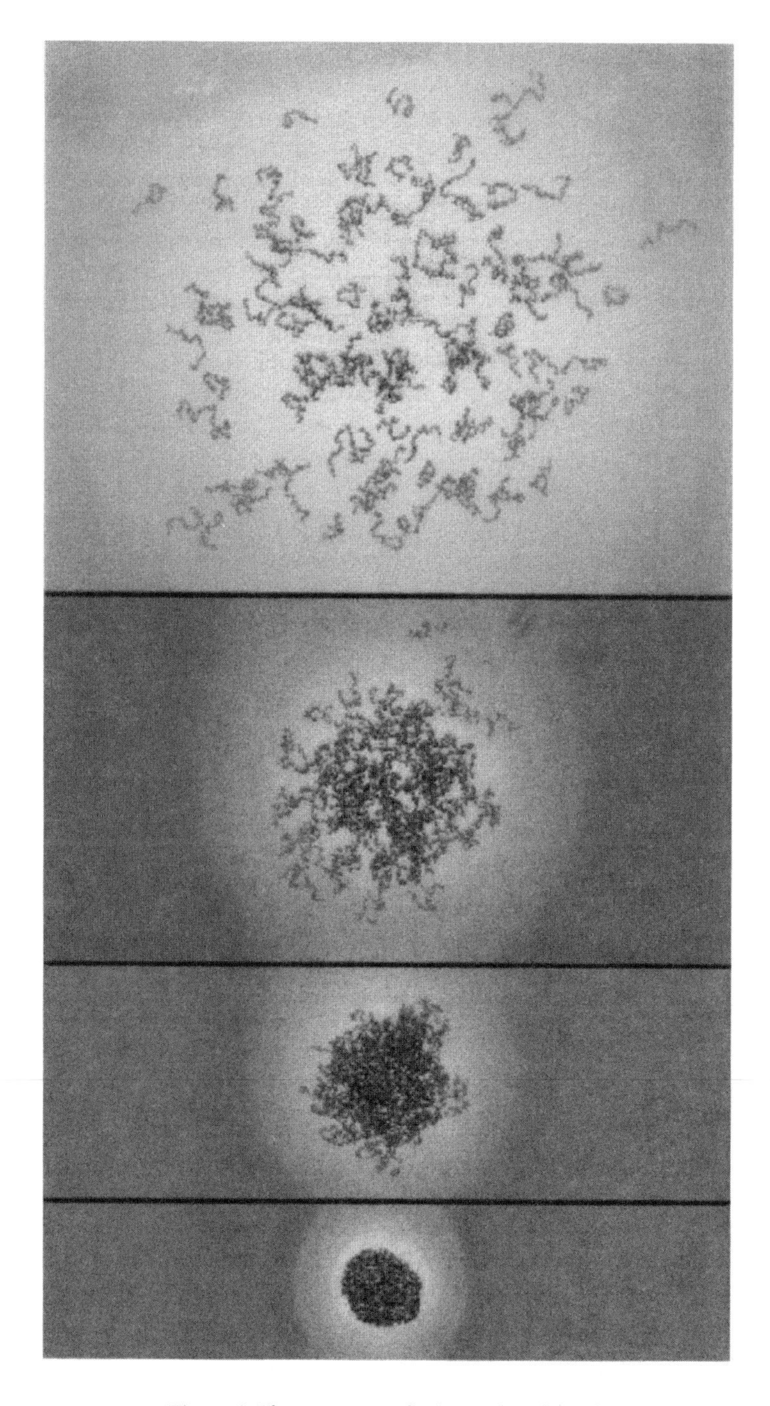

Figure 1 Time sequence of polymer drop formation

Owing to the interaction between the electron and the substrate, the effective mass of the electron in a semiconductor quantum dot is less than 10% of the actual mass of the electron.[4] We do not know the extent to which this substrate-electron interaction changes the electron's effective mass in the polymer quantum drop. In any event, the effective mass of the electron will be a strong function of the material that composes the quantum drop. Hence, in developing the model Hamiltonian, we have chosen to take the effective mass of the electron on our polymer quantum drops to be the actual mass of the electron. A second concern that must be addressed in semiconductor quantum dots is the exchange interaction among electrons confined in the same potential[14-16]. In a polymer quantum drop, each electron is confined by its own image, so that the overlap between electron orbitals is seen to be quite small and the interaction of the electrons is appropriately modeled as the Coulombic interaction between electrons.

In this investigation, the model Hamiltonian is designed to describe the interaction of n electrons when the electrons are confined to the neighborhood of the surface of a sphere. We begin with the minimum energy configuration of the electrons on the sphere surface and then fix all of the electrons to the minimum energy configuration at the surface of the sphere, except for one. The lone remaining electron, the active electron, is allowed to move under the influence of the fixed electron(s). The Hamiltonian that determines the motion of the active electron is then:

$$H = \frac{\pi^2}{2} + \left[\sum_{i=1}^{N} \frac{\xi(r_i)}{r_i} - \frac{V_0}{2}(f_1 - f_2) \right] + |E| z \tag{1}$$

where π is the canonical momentum of the active electron in the presence of a magnetic field:

$$\pi = p + \frac{\vec{A}}{c} \tag{2}$$

where the vector potential A is defined as follows in the symmetric (Coulomb) gauge in terms of the magnetic field and c is the speed of light:

$$\vec{A} = \frac{1}{2}\left[\vec{B} \times \vec{r}\right] \tag{3}$$

where ξ is the effective (screened) charge of the ith electron, V_0 is the depth of the radial potential, r_i is the separation between the active electron and the ith electron and f_1 (f_2) is a switching function that turns on (off) the radial confining potential near the surface of the sphere. For the our purposes the screening of the fixed electrons will be neglected so that all the ξ's are set to 1, V_0 is chosen to be 1.0 hartree. The total switching function (f_1 - f_2) takes the form:

$$(f_1 - f_2) = \tanh\left[\gamma\left(r - R_s + \frac{\Delta}{2}\right)\right] - \tanh\left[\gamma\left(r - R_s - \frac{\Delta}{2}\right)\right] \tag{4}$$

where r is the position of the active electron, R_s is the radius of the sphere, here chosen to be 100 au, Δ is the thickness of the radial potential that confines the active charge to the sphere (Δ = 1 au) and we choose γ to be 100. The mass and charge of the electron have been set to unity as has Planck's constant divided by 2π. The motion of this Hamiltonian system is determined by solving the Hamilton's equations for the active electron.

The Poincaré surface-of-section technique is an extension of the WKB approximation for non-separable systems in higher dimensions that has the virtue of yielding exact semiclassical results. It has been shown that this technique can be used to determine, semiclassically, the energy levels of a Hamiltonian system which exhibits quasi-periodic behavior. We use the case of three or four excess electrons for illustrative purposes in this contribution. The Poincaré surface-of-section technique has been presented elsewhere.[2,3]

ELECTRONIC STRUCTURE, STARK AND MAGNETIC EFFECTS

We have applied this semiclassical quantization scheme for the case of three electrons located on the sphere. Minimum energy conditions dictates that the initial configuration is composed of the three electrons located at the vertices of an equilateral triangle. We have chosen that the plane of the equilateral triangle should correspond with the x-z plane of our calculation. The two fixed electrons are placed at the distance R_s from the center of the sphere and at their appropriate vertex positions if the active electron where set at the position (0, 0, R_s) at the "north pole" of the sphere. The active electron is then give an initial momentum in the x direction while being initially displaced in the y direction. The classical trajectories are calculated for these initial conditions for a number of energies and the resulting table of actions and energies are analyzed to find the best fit to the action quantization conditions. Table 1 shows the result of these calculations as a table of eigenenergies for the case of three electrons on a sphere. Figure 2 shows electron trajectories on a quantum drop with four excess electrons, the electron orbits are shown in red and the polymer nanoparticle is shown in blue.

Table 1. The four lowest energy levels for a quantum drop with three electrons.

Energy Level	Energy (au)
1	-0.9883
2	-0.9880
3	-0.9876
4	-0.9871

We have done a similar calculation for the case of a four electron quantum drop in the presence of a magnetic field. The field strength that we chose is on the higher range of currently accessible fields and was chosen only to illustrate the sort of effects that a magnetic field can be expected to have on the electronic energy levels of quantum drops. Table 2 shows the lowest four pairs of energy levels of the four electron quantum drop in the presence of a 235 Tesla magnetic field. The splitting of energy levels arises from the fact that the electron can circle the "north pole" in either a clockwise or counterclockwise direction thereby giving rise to angular momentum states that either align with the direction of the magnetic field or opposing the field. This behavior is characteristic of what is expected to the interaction of the electron spin with the magnetic field. The effect on the electronic trajectories due to the external magnetic field is shown in Figures 3 and 4. Figure 3 shows a top view of the four electron quantum drop with the magnetic field oriented along the z-axis which is shown in yellow and figure 4 shows the bottom view of the same drop.

Table 3 shows the lowest four energy levels for a four electron quantum drop in the presence of a very strong electric field. Again, the field strength has been chosen only to illustrate the Stark effect in the quantum drop system. The electric field that we are considering here is about 5000 KV/m and it is seen to cause shifting of the energy levels with respect to the field free case. Figure 5 depicts the four electron quantum drop again, this time in the presence of the very strong electric field which is oriented along the z-axis again shown in yellow.

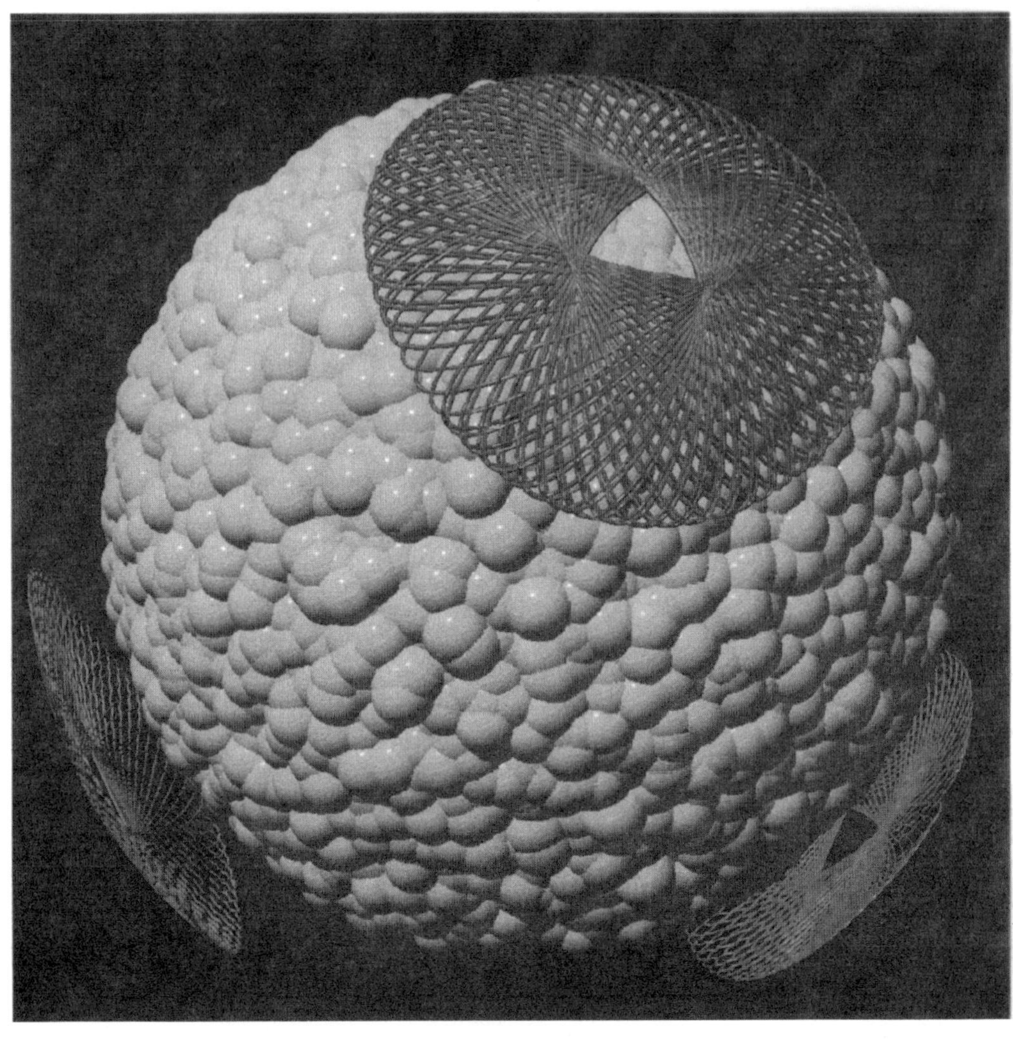

Figure 2 Electron trajectories on a ~10 nm quantum drop with four electrons in the case of no external field

Table 2. The four lowest pairs of energy levels for a quantum drop with four electrons in a 235 Tesla magnetic field.

Level	Energy (au)
1	-0.9806
2	-0.9801
3	-0.9796
4	-0.9790

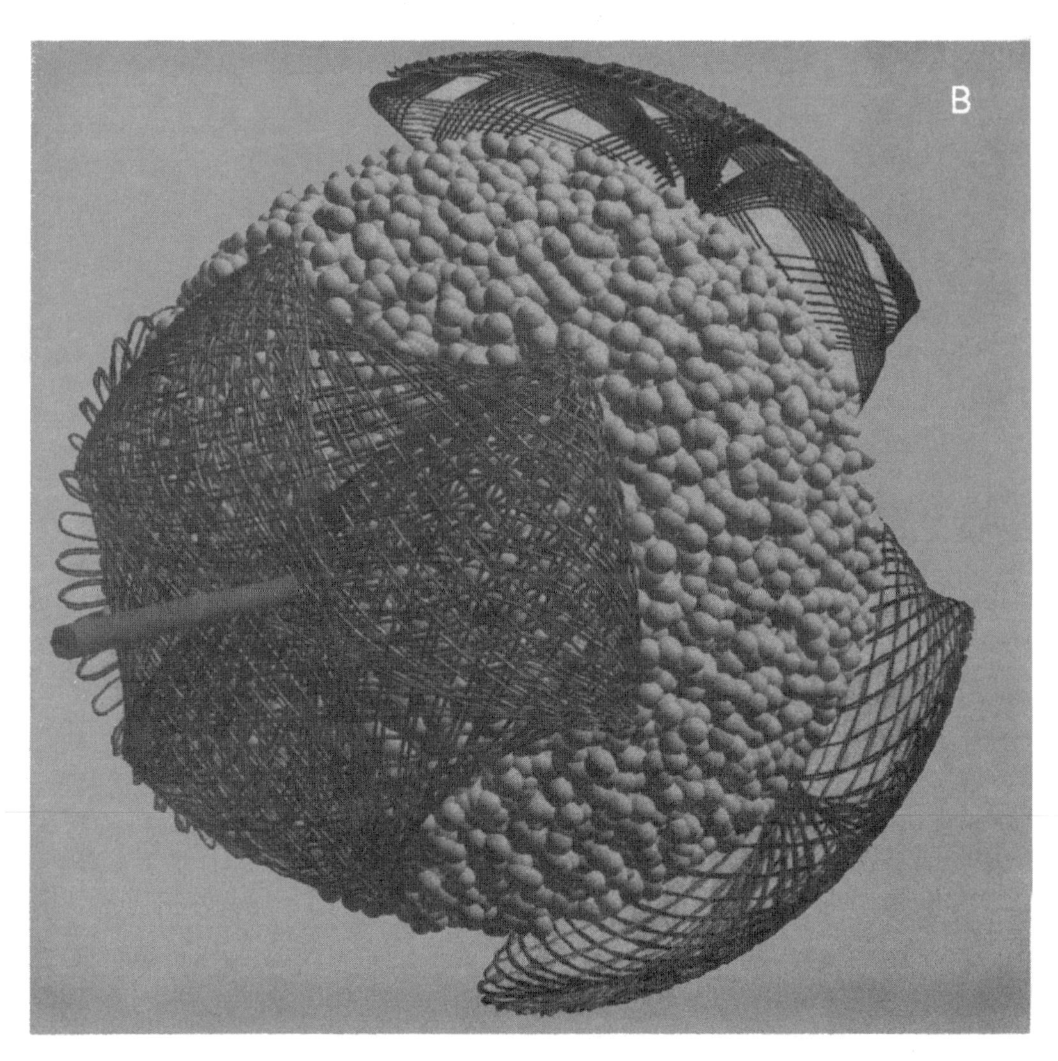

Figure 3 Top view (North Pole) of electron trajectories on a ~10 nm quantum drop with four electrons in a 235 Tesla magnetic field

Table 3. The three lowest energy levels for a quantum drop with four electrons in a ~5000 KV/m electric field compared to the field-free levels

Level	Energy (au)
1	-0.9792
2	-0.9782
3	-0.9776

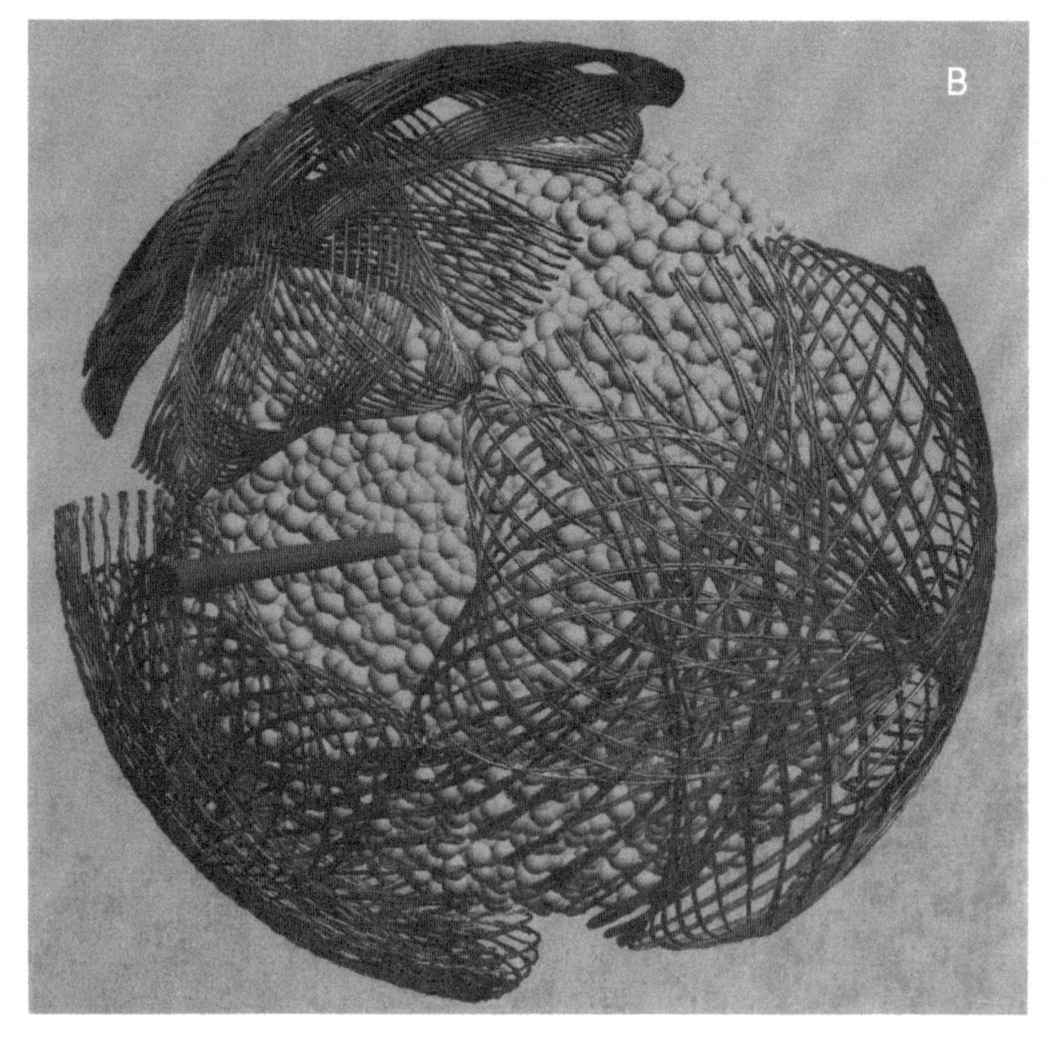

Figure 4 Bottom view (South Pole) of electron trajectories on a ~10 nm quantum drop with four electrons in a 235 Tesla magnetic field

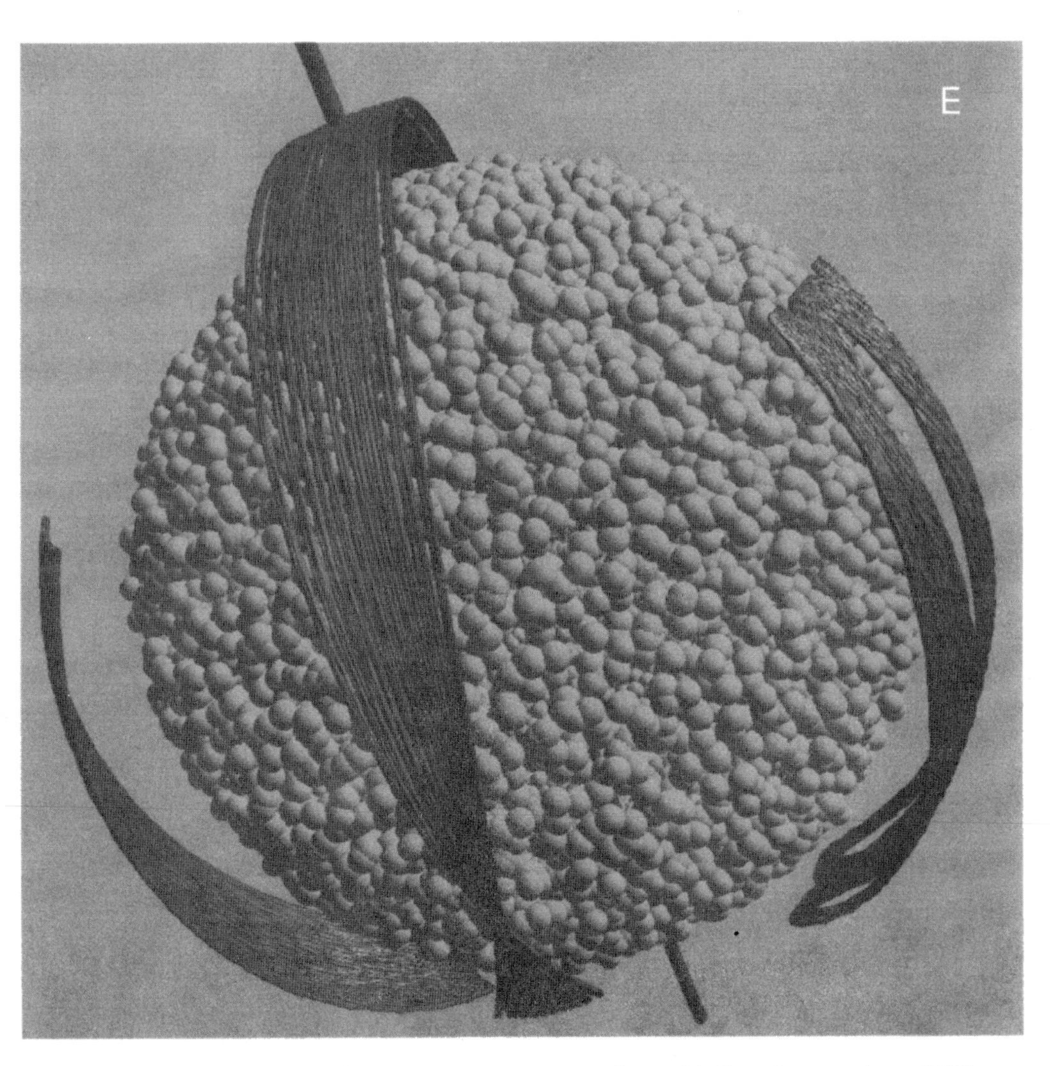

Figure 5 Electron trajectories on a ~10 nm quantum drop with four electrons in a ~5000 KV/m electric field

ACKNOWLEDGMENT :

Research sponsored by the *Division of Materials Sciences, Office of Basic Energy Sciences, U. S. Department of Energy* under Contract DE-AC05-96OR22464 with *Lockheed Martin Energy Systems, Inc.* KR supported by an appointment jointly by *Oak Ridge National Laboratory* and the *Oak Ridge Institute for Science and Education.*

REFERENCES

1. M. D. Barnes, C-Y. Kung, B. G. Sumpter, K. Fukui, D. W. Noid and J. U. Otaigbe submitted to Science (1998).
2. K. Runge, B. G. Sumpter, D. W. Noid and M. D. Barnes, submitted to J. Chem. Phys. (1998).
3. K. Runge, B. G. Sumpter, D. W. Noid and M. D. Barnes, submitted to Chem. Phys. Lett. (1998).
4. The literature on quantum dots is quite extensive. For a recent review the reader is referred to R. C. Ashoori, Nature **379**, 413 (1996), and references therein.
5. J. Blaschke and M. Brack, Phys. Rev. A **56**, 182 (1997).
6. M. Brack, J. Blaschke, S. C. Creagh, A. G. Magner, P. Meier and S. M. Reimann, Z. Phys. D **40**, 276 (1997).
7. S. Lüthi, T. Vancura, K. Ensslin, R. Schuster, G. Böhm and W. Klein, Phys. Rev. B **55**, 13088 (1997).
8. V. Zozoulenko, R. Schuster, K. -F. Berggren and K. Ensslin, Phys. Rev. B **55**, R10209 (1997).
9. Cheung, M. F. Choi and P. M. Hui, Solid State Comm. **103**, 357 (1997).
10. S. Child, *Semiclassical Mechanics with Molecular Applications* (Oxford University Press, New York, 1991).
11. D.W. Noid and R. A. Marcus, J. Chem. Phys. **62**, 2119 (1975).
12. D. W. Noid, M. L. Koszykowski and R. A. Marcus, J. Chem. Phys. **73**, 391 (1980).
13. D. W. Noid, S. K. Knudson and B. G. Sumpter, Comp. Phys. Comm. **51**, 11 (1988).
14. D. Pfannkuche, V. Gudmundsson and P. A. Maksym, Phys. Rev. B **47**, 2244 (1993).
15. P. A. Maksym, Europhys. Lett. **31**, 405 (1995).
16. B. Chengguang, R. Wenying and L. Youyang, Phys. Rev. B **53**, 10820 (1996).

SIMULATIONS OF THIN FILMS AND FIBERS OF AMORPHOUS POLYMERS

Visit Vao-soongnern,[1] Pemra Doruker,[2] and Wayne L. Mattice[1]

[1]Department of Polymer Science
The University of Akron
Akron, OH 44325-3909

[2]Department of Chemical Engineering and Polymer Research Center
Bogazici University
Istanbul, Turkey

INTRODUCTION

The small systems that are important for nanotechnology may not be "small" when viewed from the standpoint of simulations performed using models expressed at fully atomistic detail. This problem can be especially severe when the system contains amorphous polymers, because of the large range of distance and time scales that describe the relaxation of these systems. The efficiency of the simulation can be improved by resorting to a coarse-grained model for the chains, but often at the expense of an unambiguous identification of the model with any particular real polymer. This difficulty has prompted recent interest in the design, performance, and analysis of simulations that bridge representations of a single system that differ in structural detail.[1] In the present context, we require a method where there is a two-way (reversible) path connecting a coarse-grained and fully atomistic description of the same system

This chapter will review the generation of free-standing thin films and fibers with a coarse-grained simulation method on a high coordination lattice,[2,3] performed in a manner that allows accurate reverse-mapping of individual replicas to a fully atomistic representation in continuous space.[4] After describing some of the properties of these films and fibers, we will present some new information about the limits on the stability of the models of these nanofibers.

METHOD

The method adopted here uses a sparsely occupied high coordination ($10i^2+2$ sites in shell i) lattice for the coarse-grained representation of the system. This lattice is obtained by deletion of every second site from a diamond lattice.[5] In the first generation of applications of the simulation to saturated hydrocarbon melts, each occupied bead on the lattice represents a -CH_2CH_2- unit (for simulations of polyethylene[2-4,6,7]) or a -$CH_2CH(CH_3)$- unit (for simulations of atactic, isotactic, or syndiotactic polypropylene[8-10]). The step length on the lattice, 0.25 nm, is defined by the length of the C–C bond and the tetrahedral angle. The bulk density of a typical polyethylene melt is achieved with occupancy of about 1/6 of the sites on this lattice. A newer second generation of the method, in which each bead in a simulation of polyethylene represents a $CH_2CH_2CH_2CH_2$ unit, has been developed recently.[11] With the second generation of the simulation, bulk density for a polyethylene melt is achieved with occupancy of only 1/12 of the sites on the high coordination lattice. The computational efficiency of the simulation benefits from the use of a sparsely occupied lattice. The results presented here will focus on amorphous polyethylene as the material, using the first-generation method.

The Hamiltonian contains two parts. The short-range intramolecular contribution comes from the mapping of a classic rotational isomeric state model[12,13] for the real chain onto the discrete space available to the coarse-grained chain on the sparsely occupied high coordination lattice.[2,8] Specific examples that have been used in the simulations of melts of polymeric hydrocarbons are three-state rotational isomeric state models for polyethylene[14] and polypropylene.[15] The long-range and intermolecular interactions are handled by invoking self- and mutual exclusion, along with a discretization into interaction energies for successive shells (u_i, $i = 1, 2, \ldots$) of a continuous potential energy function,[3] such as a Lennard-Jones potential energy function, that describes the pair-wise interaction of small molecules[16,17] representative of the collection of atoms assigned to each bead on the high coordination lattice. For example, adoption of a Lennard-Jones potential energy function with $\varepsilon/k_B = 205K$ and $\sigma = 0.44$ nm implies that simulations of polyethylene at 443K should use $u_1 \ldots u_5$ of 14.2, 0.429, -0.698, -0.172, and –0.045 kJ/mol, respectively.[3] The first shell is strongly repulsive because it covers distances smaller than σ. The second shell is much less repulsive because it covers the distance at which the Lennard-Jones potential energy function changes sign. The major attraction appears in the third shell. If the system in the simulation is to be cohesive, at least three shells must be retained in the evaluation of the intermolecular interactions.

The simulation of polyethylene melts proceeds by random jumps (of length 0.25 nm) by individual beads to unoccupied nearest-neighbor sites on the high coordination lattice, with retention of all connections to bonded beads. These single-bead moves correspond to a variety of local moves in the underlying fully atomistic model that change the coordinates of 2 or 3 carbon atoms.[4] The acceptance or rejection of proposed moves is via the customary Metropolis criteria.[18]

When the method was applied to one-component[9] and two-component[10] melts of polypropylene chains of specific stereochemical sequence, reptation was included along with the single-bead moves, in order to achieve equilibration of the melts on an acceptable time scale. The polypropylenes (especially syndiotactic polypropylene) equilibrate slowly if the simulation uses single bead moves only.

CONSTRUCTION AND PROPERTIES OF MODELS OF THIN FILMS

Construction of the thin films commences with an equilibrated model of the melt at bulk density, contained in a three-dimensional box of dimensions $L_xL_yL_z$, measured along the three axes of the periodic cell. The angle between any two axes is 60°. Periodic boundary conditions are applied in all directions. The number and degree of polymerization of the parent chains are chosen so that the system will have bulk density, ρ_{bulk}, at the temperature of the simulation. The construction, equilibration, analysis, and reverse mapping of these models of polyethylene melts have been reviewed recently,[1,11] and will receive no additional mention here.

The equilibrated model of the melt is converted to a model of the free-standing thin film using the approach described by Misra *et al.*[19] The value of L_z is increased sufficiently so that a parent chain cannot interact with its image along the z direction. When the perturbed system is re-equilibrated, with the same procedure employed for the initial equilibration of the bulk, it now settles down into a free-standing thin film, with both surfaces exposed to a vacuum.[20] If the initial model was large enough, the free-standing film retains ρ_{bulk} in its interior. Films have been constructed and analyzed with thickness up to 12 nm. The density profiles, $\rho(z)$, near both surfaces are described by a hyperbolic tangent function.

$$\rho(z) = \frac{\rho_{\mathrm{bulk}}}{2}\left[1 - \tanh\left(\frac{z}{\xi}\right)\right] \tag{1}$$

The width parameter, ξ, has a value close to 0.5–0.6 nm, corresponding to a surface region of thickness 1.0–1.2 nm.[21] A larger thickness for the surface region is obtained if it is defined in terms of a property of the entire chain, such as the distribution of the centers of mass of the chains.[21] These results are similar to the ones obtained earlier in thin films of atactic polypropylene that were constructed by a different method.[22,23]

The anisotropy of the local environment is assessed using an order parameter S, defined using the angle, θ_z, between the z axis and a bond in the coarse-grained representation of the system.

$$S = \frac{1}{2}\left(3\left\langle\cos^2\theta_z\right\rangle - 1\right) \tag{2}$$

This order parameter is applied to individual bonds of length 0.25 nm in the coarse-grained representation (which become chord vectors in the fully atomistic representation of the same system).[20] The order parameter shows that the local environment is isotropic in the middle of the film (assuming the film is sufficiently thick), but the local environment becomes anisotropic near the surfaces. The nature of the anisotropy at the surface depends on whether the chains are linear or cyclic. For cyclic chains, S becomes negative near the surface, and remains negative, due to the tendency for internal bonds in the surface region to be oriented parallel to the surface.[21] For linear chains, S initially becomes negative as one approaches the surface from the interior, but S eventually turns positive when the density is very small, due to the tendency for the ends to be segregated at the surface, with an orientation perpendicular to the surface.[20]

The contribution made by the internal energy to the surface energy, γ, is in the range 21–22 erg/cm^2.[20,21] The surface energy is dominated by the contributions from the intermolecular interactions. The short-range intramolecular interactions make a small contribution, which tends to oppose the effect of the intermolecular interactions on γ. This result implies that bonds in the low density surface region can achieve a somewhat greater local intramolecular relaxation than is possible for bonds in the bulk region, as was observed previously by Misra *et al.*[19]

There is greater mobility in the surface region of free-standing thin films than in the bulk, both at the level of individual beads and at the level of the translational diffusion of the center of mass of chains of $C_{100}H_{202}$. This increase in mobility in the surface region is also seen for free-standing thin films that are composed of cyclic chains, $C_{100}H_{200}$. Therefore the increased mobility is attributed to the lower density in the surface region, and not to the preferential segregation of the ends of the linear chains in the surface region.[24]

The computational efficiency of the simulation with the coarse-grained model permits the study of the process by which cohesion of two thin films is obtained with *n*-alkanes.[25] Three different time scales are observed for healing of the density profile at the initial interface between the two films, redistribution of chain ends, and complete intermixing of the chains, with the time scale increasing in the order stated.

The behavior described in this section is observed in the simulations of free-standing thin films. Important changes in the static and dynamic properties can occur when the thin film interacts with a single solid wall[26] or is confined in a narrow slit between two parallel solid walls.[27]

CONSTRUCTION AND PROPERTIES OF MODELS OF THIN FIBERS

The construction of the model of a nanofiber[28] commences with an equilibrated model for a free-standing thin film. The film is continuous along the x and y axes, and it is exposed to a vacuum along the z axis. The periodicity in one of the directions along which the thin films is continuous, say the y direction, is increased by a large amount, so that the parent chains can no longer interact with their images along this direction. After a new equilibration, the model settles down into a thin fiber that is oriented along the x axis, and exposed to a vacuum along the y and z axes.

Cross-sections of the fibers perpendicular to the fiber axis are nearly circular. Models have been constructed with cross-sections that have diameters of 7–8 nm. They are comparable in thickness with some of the nanofibers prepared by the electrospinning technique, which can be as thin as 3 nm (D. H. Reneker, personal communication). If the fiber has a thickness greater than about 4 nm, it recovers bulk density in its interior. Radial density profiles perpendicular to the fiber axis can be fitted to a hyperbolic tangent function, Equation (1). For fibers with diameters in the range 5–8 nm, the correlation lengths, ξ, are about 0.6 nm, which is close to the value obtained with the models of the free-standing thin films. The end beads are enriched in the surface region, as was also the case with the free-standing thin films. The anisotropy of the chord vectors, as assessed by the order parameter, S, is also similar to the result obtained with the free-standing thin films.

Surface energies, γ, are not easily assessed for the models of the fibers. The excess energy associated with the surface is easily evaluated, but there is an ambiguity in the definition of the surface area. For the free-standing thin films, the surface area for that portion of the film in the periodic box is $2L_xL_y$, which is well defined. In contrast, the surface area of that portion of the fiber in the periodic box is $\pi d_{yz}L_x$, where d_{yz} is the diameter in the direction perpendicular to the fiber axis. This diameter is well-defined only when the density profile normal to the fiber axis is a step function. The actual radial density profile is instead described by Equation (1), with ξ close to 0.6 nm. After allowing for the ambiguity in the definition of the surface area for the fiber, it appears that the values of γ may be similar in the thin films and in the fibers.

For amorphous fibers composed of $C_{100}H_{202}$, and with thicknesses of 5–8 nm, the anisotropies of the individual chains, as measured by the principal moments of the radius of gyration tensor, are comparable with those expected for chains in the bulk. However, the radius of gyration tensors tend to be oriented within the fiber, with the nature of the orientation depending on the distance of the center of mass from the axis of the fiber. Chains with their centers of mass close to the fiber axis tend to have the largest component of their radius of gyration tensor aligned with the fiber axis.

STABILITY OF THE MODEL OF THE NANOFIBER

In this section we present new results on the stability of the models of the nanofibers. Destabilization of the model can be achieved in several ways, which include an increase in the length of the periodicity, L_x, along the fiber axis, a decrease in the number of independent parent chains, or a decrease in the degree of polymerization of the parent chains. Each change causes the fiber to become thinner. At some point the fiber breaks up into individual droplets, because the spherical shape will minimize the ratio of surface area to volume in the system.

Increase in the Length of the Periodicity along the Fiber Axis

Two different nanofibers of $C_{100}H_{202}$ were studied. One system (f36) contained 36 independent parent chains, and the other system (f72) contained 72 independent parent chains, both in boxes in which L_x was initially 5.25 nm. The simulation protocol for collapse of the fiber consisted of a cycle of increasing L_x by 0.25 nm, followed by relaxation for 10^5 Monte Carlo steps at a temperature of 509K. This cycle was continued until collapse was observed. The smaller system with 36 independent parent chains collapsed to droplets when L_x reached 10 nm. The larger system, with 72 independent parent chains, did not collapse even when L_x exceeded the length of the fully extended chain, which is 12.5 nm.

Figure 1 depicts radial density profiles, measured normal to the fiber axis, for the f36 system at several values of L_x. This figure suggests two different methods for detecting the onset of the breakup of the fiber. One method uses the density at the core of the fiber. In the initial system, with $L_x = 5.25$ nm, there is an extensive region inside the fiber where the density is constant at about 0.72–0.73 g/cm^3, which is a reasonable value for ρ_{bulk} at this temperature. This region of constant density is reduced in extent as L_x increases. Eventually no part of the system, measured relative to the initial fiber axis, retains a density as high as 0.7 g/cm^3. One way of defining the point at which break-up of the fiber occurs is to identify that point with the loss of the region at the core where the density is constant at ρ_{bulk}. However, this method is subject to statistical error. It relies heavily on events in a relatively small number of cells (those very close to the initial fiber axis), and therefore is subject to a high statistical uncertainty, as is evident from the scatter in the

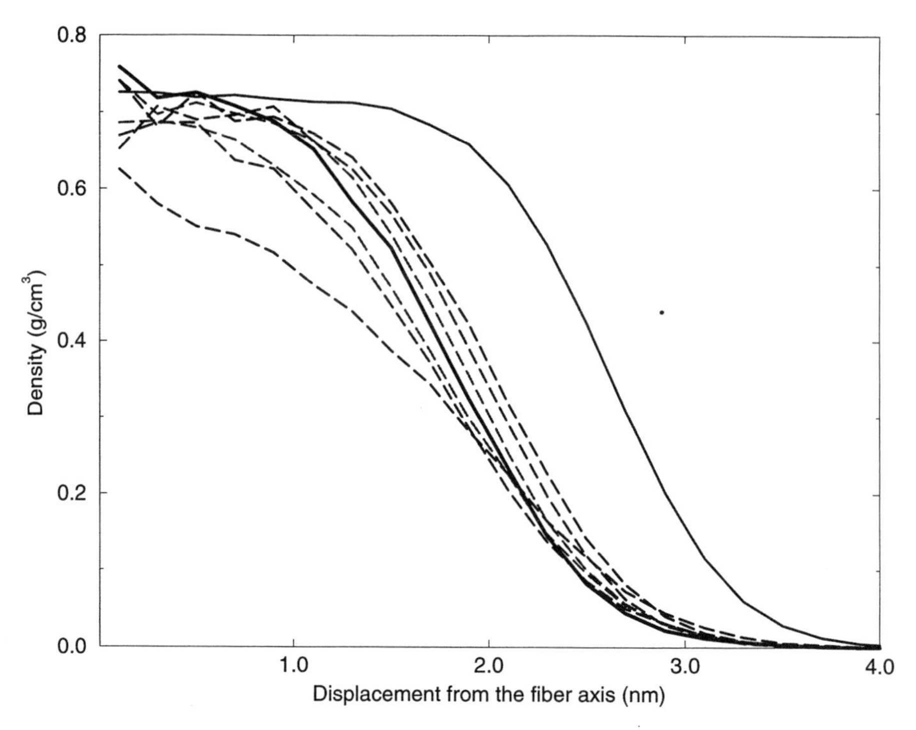

Figure 1. Radial density profiles, normal to the fiber axis, for a fiber composed of 36 independent parent chains of $C_{100}H_{202}$ at 509K. The periodicity of the simulation box along the fiber axis, L_x, is 11.5, 11.0, 10.5, 10.0, 9.5, 9.0, 8.5, and 5.25 nm, reading from left to right.

data at the extreme left hand side of Figure 1. This statistical problem can be alleviated using an alternative definition of the onset of disruption, using results derived from a larger number of occupied cells.

A larger number of cells can be employed if the focus is shifted from the inside of the fiber to the surface region, where the density is a strong function of distance from the fiber axis. With the initial structure, formed with $L_x = 5.25$ nm, the density profile in the surface region is described by the hyperbolic tangent function in Equation (1). As L_x increases, the density profiles in the surface region initially retains nearly the same shape, but eventually, as L_x continues to increase, the density profiles in the surface region become broader. This broadening shows up in the fits to Equation (1) as in increase in the correlation length, ξ. Using the size of ξ as the criterion, the breakup of the fiber treated in Figure 1 has its onset when L_x is 10.0 nm. This identification is consistent with visual inspection of fibers that have been subjected to reverse mapping, which restores the missing carbon and hydrogen atoms.

The total energy (sum of the intramolecular contribution from the rotational isomeric state model and the intermolecular contribution from the discretized Lennard-Jones potential energy function) of the f36 system initially increases as L_x increases, passes through a broad maximum when L_x is near 10 nm, and then slowly decreases. This behavior is consistent with the identification of the onset of the breakup of the fiber at L_x = 10 nm, but it is less precise than the identification using ξ, because the maximum in the total energy is very broad. The trend for the total energy is dominated by intermolecular contributions from the discretized Lennard-Jones potential energy function. There is no apparent trend in the intramolecular energies from the rotational isomeric state model.

Decrease in the Number of Independent Parent Chains

Figure 2 depicts radial density profiles for a series of simulations that initiate with the fiber composed of 76 independent parent chains of $C_{100}H_{202}$, initially in the periodic box with L_x = 5.25 nm. Several of the parent chains are removed from the system, and the perturbed system is then re-equilibrated at 509K. The fiber becomes thinner as chains are removed, but the core retains its integrity with as few as 14 independent parent chains. Disruption of the core of the fiber is readily apparent when the number of independent parent chains decreases to 9.

Decrease in the Degree of Polymerization of the Parent Chains

Figure 3 depicts radial density profiles for fibers in a series of simulations that initiate with the fiber composed of 36 independent parent chains of $C_{100}H_{202}$, in the periodic box with L_x = 5.25 nm. Several beads are removed from the ends of the parent chains, and the perturbed system is then re-equilibrated at 509K. The fiber retains its integrity when the

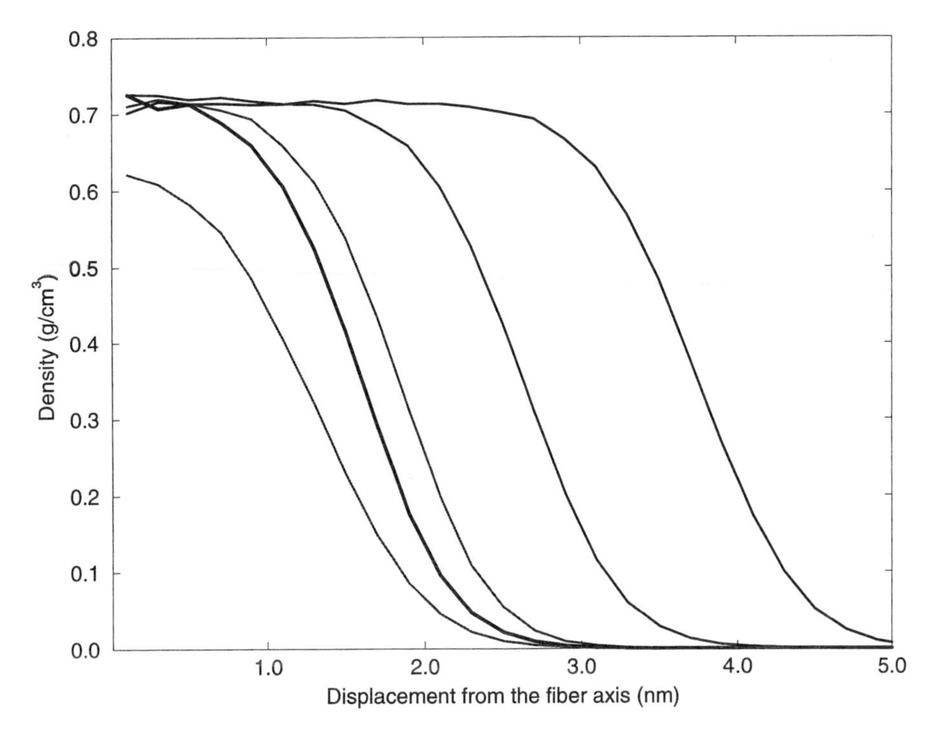

Figure 2. Radial density profiles at 509K, normal to the fiber axis, for a fiber composed of 9, 14, 18, 36, and 72 independent parent chains of $C_{100}H_{202}$, reading from left to right. The periodicity along the fiber axis, L_x, is 5.25 nm.

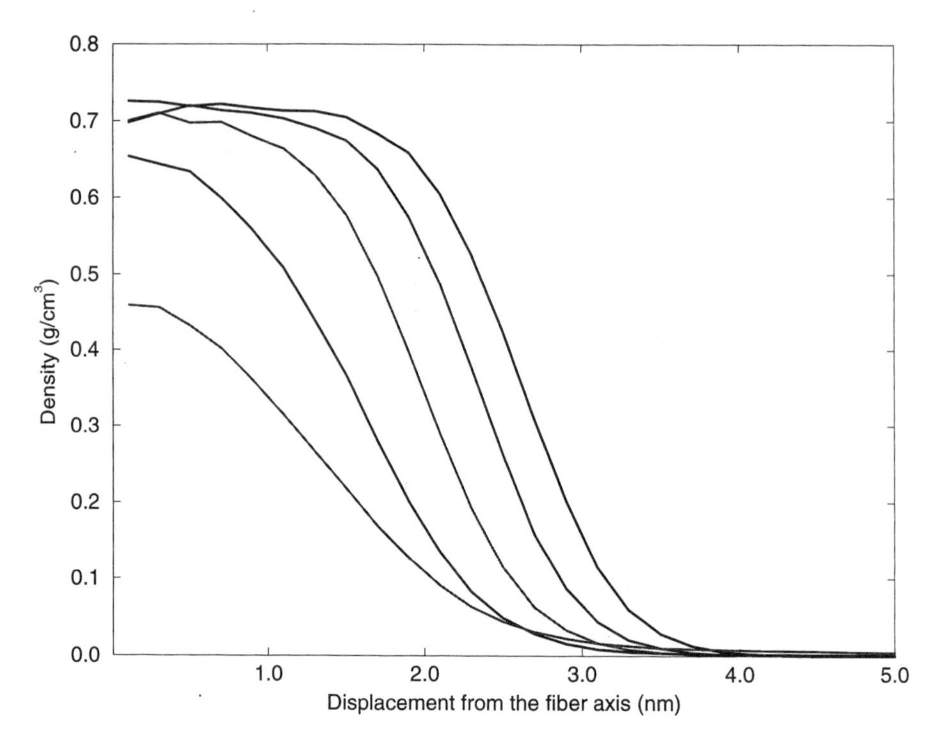

Figure 3. Radial density profiles at 509K, normal to the fiber axis, for a fiber composed of 36 independent parent chains of $C_{30}H_{62}$, $C_{40}H_{82}$, $C_{60}H_{122}$, $C_{80}H_{162}$, and $C_{100}H_{202}$, reading from left to right. The periodicity of the simulation box along the fiber axis, L_x, is 5.25 nm.

parent chains are reduced in length to $C_{80}H_{162}$, but disruption is apparent when the parent chains are as short as $C_{40}H_{82}$. The onset of disruption occurs when the chains are reduced in length to approximately $C_{60}H_{122}$.

CONCLUSION

Models for nanofibers of amorphous polymers can be constructed by manipulation of the periodic boundary conditions commonly employed in the simulations of polymer melts. After a strong extension of the periodicity in one direction, equilibration causes the model to form a free-standing thin film. Repetition of this process with the boundary in a different direction converts the free-standing thin film into a thin fiber, exposed to a vacuum. For a successful construction of the model for the fiber, the system must be sufficiently robust, in terms of the degree of polymerization and number of the independent parent chains. The length of the periodic box along the fiber axis can exceed the length of a fully extended parent chain if a sufficiently large number of independent parent chains is employed in the simulation. The correlation length, ξ, in the fit of the radial density profile to a hyperbolic tangent function is a useful indicator of the integrity of the model for the thin fiber.

ACKNOWLEDGMENTS

This work was supported by the NSF/EPIC Center for Molecular and Microstructure of Composites and by NSF DMR 9844069. V.V. expresses thanks to The Royal Thai Government for the Development and Promotion of Science and Technology Talents Project (DPST) Scholarship during his graduate study at The University of Akron.

REFERENCES

1. J. Baschnagel, K. Binder, P. Doruker, A. A. Gusev, O. Hahn, K. Kremer, W. L. Mattice, F. Müller-Plathe, M. Murat, W. Paul, S. Santos, U. W. Suter, and V. Tries, Bridging the gap between atomistic and coarse-grained models of polymers: Status and perspectives, *Adv. Polym. Sci.*, in press.
2. R. F. Rapold and W. L. Mattice, Introduction of short and long range energies to simulate real chains on the 2nnd lattice, *Macromolecules* 29:2457 (1996).
3. J. Cho and W. L. Mattice, Estimation of long-range interaction in coarse-grained rotational isomeric state polyethylene chains on a high coordination lattice, *Macromolecules* 30:637 (1997).
4. P. Doruker and W. L. Mattice, Reverse mapping of coarse grained polyethylene chains from the second nearest neighbor diamond lattice to an atomistic model in continuous space, *Macromolecules* 30:5520 (1997).
5. R. F. Rapold and W. L. Mattice, New high-coordination lattice model for rotational isomeric state polymer chains, *J. Chem. Soc., Faraday Trans.* 91:2435 (1995).
6. P. Doruker and W. L. Mattice, Dynamics of bulk polyethylene on a high coordination lattice, *Macromol. Symp.* 133:47 (1998).
7. R. Ozisik, P. Doruker, E. D. von Meerwall, and W. L. Mattice, Translational diffusion in Monte Carlo simulations of polymer melts: Center of mass displacement *vs.* integrated velocity autocorrelation function, *Comput. Theor. Polym. Sci.*, submitted.
8. T. Haliloglu and W. L. Mattice, Mapping of rotational isomeric state chains with asymmetric torsional potential energy functions on a high coordination lattice: Application to polypropylene, *J. Chem. Phys.* 108:6989 (1998).
9. T. Haliloglu, J. Cho, and W. L. Mattice, Simulations of rotational isomeric state models for poly(propylene) melts on a high coordination lattice, *Macromol. Theory Simul.* 7:613 (1998).
10. T. Haliloglu and W. L. Mattice, Detection of the onset of demixing in simulations of polypropylene melts in which the chains differ only in stereochemical composition, *J. Chem. Phys.* 111:4327 (1999).
11. P. Doruker and W. L. Mattice, A second generation of mapping/reverse mapping of coarse-grained and fully atomistic models of polymer melts, *Macromol. Theory Simul.* 8:463 (1999).
12. P. J. Flory. *Statistical Mechanics of Chain Molecules*, Wiley, New York (1969).
13. W. L. Mattice and U. W. Suter. *Conformational Properties of Large Molecules. The Rotational Isomeric State Model in Macromolecular Systems*, Wiley, New York (1994).
14. A. Abe, R. L. Jernigan, and W. L. Mattice, Conformational energies of n-alkanes and the random configuration of higher homologues including polymethylene, *J. Am. Chem. Soc.* 88:631 (1966).
15. U. W. Suter, S. Pucci, and P. Pino, The epimerization of 2,4,6,8-tetramethylnonane and 2,4,6,8,10-pentamethylundecane, low molecular weight model compounds of polypropylene, *J. Am. Chem. Soc.* 97:1018 (1975).
16. J. O. Hirschfelder, C. F. Curtiss, and R. B. Bird. *Molecular Theory of Gases and Liquids*, Wiley, New York (1954).
17. J. Prausnitz. *Molecular Thermodynamics of Fluid-Phase Equilibrium*, Prentice Hall, Inc., Englewood Cliffs, New Jersey (1986).
18. N. Metropolis, A. W. Rosenbluth, A. H. Rosenbluth, A. H. Teller, and E. Teller, Equation of state calculations by fast computing machines, *J. Chem. Phys.* 21:1087 (1953).
19. S. Misra, P. D. Fleming III, and W. L. Mattice, Structure and energy of thin films of poly(1,4-cis-butadiene): A new atomistic approach, *J. Comput.-Aided Mater. Des.* 2:101 (1995).
20. P. Doruker and W. L. Mattice, Simulation of polyethylene thin films on a high coordination lattice, *Macromolecules* 31:1418 (1998).
21. P. Doruker and W. L. Mattice, Segregation of chain ends is a weak contributor to increased mobility at free polymer surfaces, *J. Phys. Chem. B* 103:178 (1999).

22. K. F. Mansfield and D. N. Theodorou, Atomistic simulation of a glassy polymer surface, *Macromolecules* 23:4430 (1990).
23. K. F. Mansfield and D. N. Theodorou, Molecular dynamics simulation of a glassy polymer surface, *Macromolecules* 24:6283 (1991).
24. P. Doruker and W. L. Mattice, Mobility of the surface and interior of thin films composed of amorphous polyethylene, *Macromolecules* 32:194 (1999).
25. J. H. Jang and W. L. Mattice, Time scales for three processes in the interdiffusion across interfaces, *Polymer* 40:1911 (1999).
26. J. H. Jang and W. L. Mattice, The effect of solid wall interactions on an amorphous polyethylene thin film, using a Monte Carlo simulation on a high coordination lattice, *Polymer* 40:4685 (1999).
27. J. H. Jang and W. L. Mattice, A Monte Carlo simulation for the effect of compression of an amorphous polyethylene melt in very thin confined geometry, *Macromolecules* 33:1467 (2000).
28. V. Vao-soongnern, P. Doruker, and W. L. Mattice, Simulation of an amorphous polyethylene nanofiber on a high coordination lattice, *Macromol. Theory Simul.* 9:1 (2000).

IDENTIFYING AND DESIGNING OF CALCIUM BINDING SITES IN PROTEINS BY COMPUTATIONAL ALGORITHM

Wei Yang[1], Hsiau-Wei Lee[2], Michelle Pu[2], Homme Hellinga[3], and Jenny J. Yang[2]

[1]Department of Biology
[2]Department of Chemistry, Center for Drug Design
Georgia State University
[3]Department of Biochemistry, Medical Center, Duke University

INTRODUCTION

Calcium is an essential component in the biomineralization of teeth, bones, and shells, as well as a second messenger regulating cellular processes such as cell division and growth, secretion, ion transport, and muscle contraction (Da Silva & Williams, 1991; Linse & Forsen, 1995). Calcium ions regulate many different biological processes by binding to proteins with different affinity. The binding of calcium to proteins leads to an increase of stability and changes in conformations of the calcium binding proteins. There are many calcium-binding proteins located in different cellular compartments, such as extracellular, intracellular, and the nucleus (Fig. 1). The mediation and regulation of Ca(II)-dependent functions are fulfilled by the change of calcium concentrations in the different cellular environments. The understanding of the molecular basis of diseases caused by the overloading of calcium, the disruption of calcium binding sites of proteins, and the mechanism of calcium-modulated signal transduction requires the establishment of the principles for the affinity of calcium binding proteins (Kawaski & Kretsinger, 1995; Schafter & Heizman, 1996).

The rational design of novel proteins has been shown to be a powerful approach for the identification of the key factors involved in protein functions, to study structure and protein folding, and to establish principles for the prediction of the function of the proteins (Bryson et al., 1995; Hellinga, 1998). In addition, protein design is a way to construct new biomaterials, sensors, catalysts and pharmaceutical drugs. Rapid progress has been made in designing metal ions such as zinc, cobalt and nickel in proteins (Regan, 1995; Shi et al., 1996; Lu & Valentine, 1997). However, little progress has been made in the design of Ca(II) binding proteins. So far, only a handful of designed Ca(II) binding studies have been published. All of them are based on the transfer of a Ca(II) binding loop into a structural analog, such as the surface-exposed Ca(II) binding loop from thermolysin into neutral protease (Toma et al., 1991), and that from α-lactalbumin into egg white lysozyme (Aramini et al., 1992; Regan, 1993). The *de novo* design of calcium binding sites, however, has not been achieved. Designing calcium-binding sites in non-calcium-binding proteins will deepen our understanding of the calcium-protein interaction without interfering with the metal-metal interaction. This helps us to understand the key determinants for calcium-binding affinity, and brings us closer to the design of novel calcium-modulated proteins with specifically desired functions.

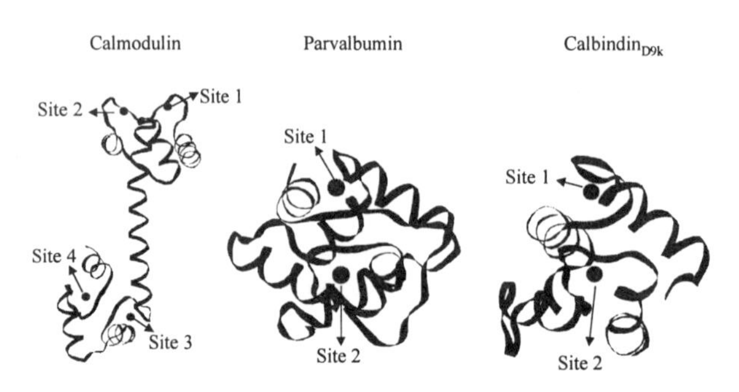

Figure 1. Three-dimensional structures of the intracellular EF-hand calcium-binding proteins calmodulin (3cln), parvalbumin (5cpv), and calbindin$_{D9k}$ (4icb).

One of the major barriers for the design of calcium binding sites in proteins comes from the complexity and irregularity of calcium binding sites. Detailed surveys (Strynadka & James, 1989; da Silva & Williams, 1991; Glusker, 1991; Falke et al., 1994; Kawasaki & Kretsinger, 1995; Nelson & Chazin, 1998) of more than one hundred known Ca(II)-binding proteins and several hundreds of Ca(II)-small molecule complexes have revealed that almost all of the ligands at Ca(II) binding sites in proteins are oxygen atoms (Fig. 2). The side chains of Asp, Glu, and Asn, the carbonyl groups from the main chain, and water from solvents (Glusker, 1991; Falke et al., 1994) all provide oxygen atoms to chelate calcium. The deviations in the Ca-O bond length are from 2.2 to 2.8 Å with a mean Ca(II)-ligand distance of 2.42 Å (Glusker, 1991; Falke et al., 1994). The coordination number varies from 3 to 9 while 6 or 7 coordination atoms are commonly found in proteins. The most common calcium-binding site has a pentagonal bipyramid or distorted octahedral geometry (Fig. 2). Ca(II)-binding sites with coordination numbers of seven or eight tend to employ more bidentate and water ligands, possibly due to ligand crowding around the ions.

A class of evolutionarily related intracellular calcium binding proteins (EF-hand proteins) have been shown to have strong homology both in the protein sequence, protein frame, and calcium binding sites (Kuboniwa et al., 1995; Finn et al., 1995; Zhang et al., 1995; Nelson & Chazin, 1998) (Fig. 1). To date, there are more than 500 appearances of EF-hand proteins in the protein and gene data banks. EF-hand proteins contain a variety of subfamilies, such as parvalbumin (Parv), troponin C (TnC), calmodulin (CaM), sarcoplasmic calcium binding protein, the essential and regulatory light chains of myosin, calbindin$_{D9K}$ (CBD), the S100, and VIS subfamilies (Linse & Forsen, 1995; Kawasaki & Kretsinger, 1995). All the EF-hand motifs we know of consist of a highly conserved loop flanked by two helices (helix-loop-helix) (Kretsinger & Nockolds, 1973), which can be further divided into classic EF-hand motifs and pseudo-EF-hand motifs. The residue numbers in the calcium-binding loop are 12 and 14 for the classic EF-hand and pseudo-EF-hand motifs, respectively. For a classic EF-hand motif (Fig. 2), seven oxygen atoms from the sidechains of Asp, Asn, and Glu, the main chain, and water at the loop sequence positions of 1, 3, 5, 7, 9, and 12 coordinate the calcium ion in a pentagonal bipyramidal arrangement. In a typical geometry, position 1 of the calcium-binding loop is always Asp and the side-chain of Asp serves as a ligand on the x-axis. The –x-axis (position 9) is filled with a bridged water molecule connecting the sidechain of Asp, Ser and Asn (Fig. 2) (Kretsinger and Nockolds, 1973; Strynadka and James, 1989). Axis -z is shared by the two carboxyl oxygen atoms of a glutamate side chain at position 12 that binds in a bidentate mode to Ca(II). Glu is used predominantly (92%) as a bidentate ligand for both classic and pseudo EF-hand motifs in all intracellular calcium-binding proteins (Falke et al., 1994; Kawasaki & Kretsinger, 1995). For the calcium-binding loop of a pseudo-EF-hand motif, four oxygen atoms from the main-chain carbonyl groups at sequence positions 1, 4, 6, and 9 provide ligands for the calcium ion with a coordination geometry very similar to that of a classic EF-hand motif (Svensson et al., 1992; Linse & Forsen, 1995). Two EF-hand motifs are almost always associated in the same domain of a protein to yield highly cooperative calcium binding systems (Falke et al., 1994; Kawasaki & Kretsinger, 1995). Calmodulin, for example, contains four EF-hand calcium-binding sites in two domains (Fig. 1). $Calbindin_{D9k}$ contains a pseudo EF-hand motif for calcium binding site 1, while it contains a classic EF-hand motif for site 2. Although EF-hand proteins have shown strong homology in primary sequences and similarity in the protein frame and calcium binding sites, their cellular functions, especially the response to calcium binding, are extremely diverse. Calcium binding tightly regulates the functions of trigger-like EF-hand proteins, such as calmodulin and troponin C, since calcium binding to trigger proteins leads to large changes in their conformations and the exposure of their hydrophobic surfaces. On the other hand, calcium binding does not result in a significant conformational change of non-trigger or buffer proteins, such as $Calbindin_{D9K}$, whose functions are proposed to maintain proper cellular calcium concentrations. The origins for the different responses to calcium binding between trigger and non-trigger proteins are not clear (Nelson & Chazin, 1998). It is necessary to investigate factors contributing to both classes of EF-hand proteins.

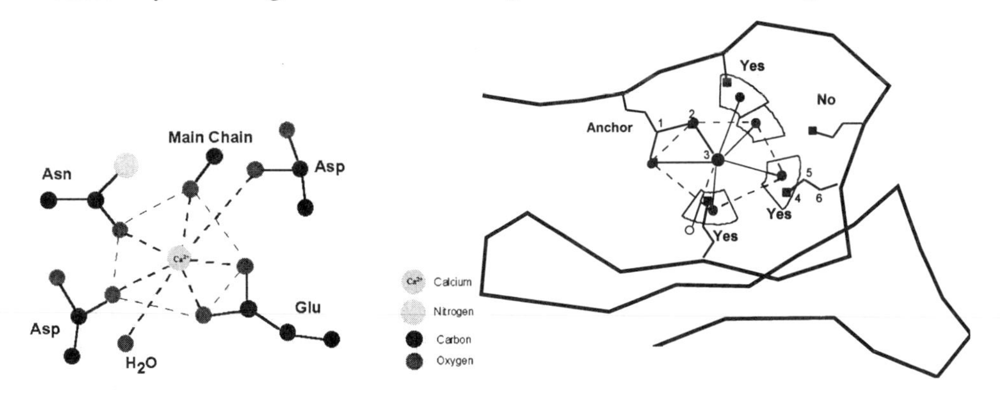

Figure 2. The calcium-binding site of calmodulin with a pentagonal bipyramidal geometry (left). Strategy of Dezymer program (right). The first residue located in the calculation (called anchor) defines the relative position of the calcium atom to the protein backbone and is used as a starting point to construct a calcium-binding site. After attaching the anchor residue to the backbone of the protein along the protein sequence, the calcium-binding geometry or positions of other ligands are then defined around the anchor.

The rational design of calcium binding sites into proteins requires several capabilities for a computer algorithm such as placing large coordination numbers (e.g. 7 and 8), incorporating different ligand types and accommodating the irregularity of metal binding sites. Several computer programs have been developed for the design of metal binding sites in proteins. Algorithms based on the analysis of hydrophobicity and charge valences of metal binding sites in protein data banks have been developed in Eisenberg's group and in Cera's lab (Yamamshita et al., 1990; Nayal & Cera, 1994). METAL SEARCH has been established for the *de novo* design of tetrahedral Zn(II) sites in proteins (Clarke & Yuan, 1995). This program has been successfully tested in the design of sites in $\alpha4$ and $\beta1$ using His and Cys as ligands (Regan & Clarke, 1990; Klemba et al., 1995).

In this work, we report our studies to establish structural parameters for identifying and designing calcium-binding proteins by the computer algorithm Dezymer. Different metal sites with low ligand numbers such as zinc fingers, blue copper, and $[Fe_4S_4]$ clusters have been successfully constructed in the hydrophobic core of *Escherichia Coli* thioredoxin by Hellinga and his colleagues (Hellinga & Richards, 1991; Hellinga, 1998). Here, we have used three EF-hand proteins, calmodulin, parvalbumin and calbindin$_{D9K}$, as model systems to test if this computer program has the ability to accommodate the intrinsic variability of ligand geometry and length of Ca(II)-ligand oxygen atom. We demonstrate that oxygen atoms from mainchain carbonyl and water can be implanted as ligands. Both classic and pseudo EF-hand binding sites in natural EF-hand proteins can be relocated using the parameters from the geometry of an ideal pentagonal bipyramid. We further demonstrate that the natural sites have the smallest deviation from the ideal pentagonal bipyramidal geometry.

METHODS

The Dezymer algorithm uses a geometric description of the ligands around a metal, the three-dimensional structure of a protein, and a library of sidechain rotamers of amino acids (or atoms from the mainchain) to identify a set of potential binding sites (Hellinga &, Richards, 1991). In principle, this program can search for any coordination number, geometry and combination of amino acid ligands based on the appropriate positioning of amino acids. The identification of potential calcium-binding sites in natural calcium-binding proteins is divided into two major steps: Site-Search and Refinement.

In the Site-Search step of this study, the high-resolution X-ray structures of calmodulin (3cln) (Babu et al., 1988), parvalbumin (5cpv) (Swain et al., 1989) and calbindin$_{D9k}$ (4icb) (Svensson et al., 1992) are used. Calcium binding sites in all three proteins are removed by deleting all the heteroatoms in these structural files including water and calcium molecules. The first residue located in the calculation (called anchor) defines the position of the calcium atom in the protein frame and is used as a starting point to construct a calcium-binding site. Since both classical and pseudo EF-hand motifs use glutamate as a bidentate ligand, Glu residue was used as an anchor for the design of all calcium binding sites. A calcium atom is placed on the same plane of two sidechain oxygen atoms and the Cδ atom with a distance between the calcium and oxygen atoms of 2.4 Å. The angle of O-Ca-O and Ca-O-Cδ are set to 53.8° and 93.5°, mimicking the natural interaction between the bidentate glutamate and calcium. This anchor residue (Glu-Ca) shares the same rotamer library of the Glu sidechain with 24 configurations (Hellinga & Richards, 1991). After attaching the anchor residue to the backbone of the protein along the protein sequence, the calcium-binding geometry or positions of other ligands are then defined around the anchor (Fig. 2). The geometry of pentagonal bipyramidal, with five oxygen atoms on the same plane and the other two above or below the plane, was used to describe all calcium-binding sites in natural EF-hand proteins. The two oxygen atoms from the anchor residue Glu are placed on the plane. One position either above or below the plane is deliberately unoccupied to allow for the oxygen atom from water to be used as a ligand. The other four ligand-positions are further defined according to the relative position of each ligand residue to the anchor residue using oxygen atoms from the sidechain oxygens of Asp and Asn and the mainchain oxygen from the protein backbone. To define the relative position between the anchor and a ligand, a total of six atoms, three from the anchor residue (atom 1, 2 and 3) and three from the residues being tested (atom 4, 5 and 6), are used. Atoms 3 and 4 are the calcium atom attached to the anchor and the oxygen atom from the ligand residue, respectively (Fig. 2). The other four atoms are Cδ and Cγ in Glu-Ca and Cγ and Cβ in Asp or Asn. The position of one potential ligand is defined by the length of atom 3_4, the angles of atoms 2_3_4 and 3_4_5, and the dihedral angles of atoms 1_2_3_4, 2_3_4_5 and 3_4_5_6. To define the position of the oxygen atom from a ligand residue in a calcium-binding site, the length of 3_4, angle of 2_3_4 and dihedral angles of 1_2_3_4 need to be described (Fig. 2). All of these parameters are given limited ranges. The other intra-residue parameters such as dihedral angles for atoms within the ligand residue are not constrained (-180° – 180°), but are restricted by the sidechain configurations of each residue type (rotamers) and the protein backbone. The complete construction of a potential site must meet two criteria. First, all the geometric definitions of the metal site must be fit. Secondly, the positions of potential ligand atoms should not overlap with the existing atoms such as the backbone of the protein or pre-located residues. If any configurations of the potential sites cannot satisfy the geometry definition or if there is clash with neighboring atoms, this potential site is rejected.

In the Refinement step, the accepted test configurations generated from the previous Site-Search step will be minimized using the Polak-Ribiere non-linear conjugate gradient algorithm (Press et al., 1988) with equation 1 (Hellinga & Richards, 1991).

$$U(P_i) = \omega_L \sum_{j=1}^{N_L} (1/\sigma^2_{Lj})(l_{ij} - L_j)^2 + \omega_\Omega \sum_{j=1}^{N_\Omega} (1/\sigma^2_{\Omega j})(\omega_{ij} - \Omega_j)^2 + \omega_\theta \sum_{j=1}^{N_\theta} (1/\sigma^2_{\theta j})(\theta_{ij} - \Theta_j)^2 \quad (1)$$

(bond lengths) (bond angles) (dihedral angles)

$$+ \omega_S \sum_{j=1}^{N_S} \sum_{k=1}^{3} (1/\sigma^2_{Sj})(\chi_{ijk} - X_{jk})^2 + \omega_{vdw} \sum_{j=1}^{N_A} \sum_{k=1, k\neq j}^{N_A} \begin{cases} d_{jk} < (r_j + r_k), (1/\sigma^2_{vdw})(d_{jk} - (r_j + r_k))^2 \\ d_{jk} > (r_j + r_k), 0 \end{cases}$$

(shift constraints) (non-bonded contacts)

where P_i is the test configuration I, ω_L, ω_Ω, ω_Θ, ω_σ and ω_{vdw} are weights for bond lengths $1 \ldots N_L$ (Hellinga & Richards, 1991), bond angles $1 \ldots N_\Omega$, position shift constraints of amino acid backbone atoms $1 \ldots N_s$, and van der Waals' contact between all atom pairs $1 \ldots N_A$ in the test configuration. σ_X are the standard deviations for parameter type X: l_j and L_j are the measured and target jth bond lengths, respectively. Similar definitions are given for bond angle ω_j and Ω_j, dihedrals θ_j and Θ_j, coordinates x_j and X_j, and the distance d_{jk} between non-bonded atoms j and k with the atomic hard-sphere radii r_i and r_j, respectively (Hellinga & Richards, 1991).

The first three items in equation 1 are the geometry deviation, such as the differences of oxygen-calcium-oxygen angle, oxygen-calcium distance and O-Ca-O-Cδ dihedral angle between the constructed site and the target site. For all of the studies here, a geometry of pentagonal bipyramidal is used to describe the target site. The distance between oxygen and calcium is 2.4 Å. 72° and 90° are used for the oxygen-calcium-oxygen angle between neighboring oxygen on the same plane, and that between oxygen atoms on the plane and off the plane, respectively. The target O-Ca-O-Cδ for the oxygen on the same plane is 0° and for that off the plane is 90°. The positions of all the atoms from sidechain and backbone of ligand residues in a constructed site are also shifted to minimize the deviation pseudo-energy. The minimized pseudo-energy U(P) gives a quantitative evaluation of the deviation between the site and target site. All the minimized sites are further filtered using the computer program, CLEAN (Yang et al., unpublished results) to remove the degenerated sites. Sites with the lowest U(p) number in the degenerated sites will be further analyzed.

RESULTS AND DISCUSSION

Description of the Geometry of Natural EF-hand sites

To seek a common parameter set that can be used to identify the natural calcium binding sites, an ideal pentagonal bipyramidal geometry with bond length of Ca-O 2.4 Å is used. As listed in Table 1, different types of oxygen ligands of natural calcium binding sites in calmodulin, parvalbumin and calbindin$_{D9k}$ have very similar Ca-O distances with an average value of 2.4 ± 0.4 Å (Table 1). The bond lengths Ca-O therefore are constrained within a relatively narrow range from 1.5-3.5 (2.5 ± 1.0) Å for design purposes. On the other hand, the angles of O-Ca-O differ largely in the natural calcium-binding sites (Table 2). In reference to the Oε atom of carboxylate bidentate Glu, the O-Ca-O angle for two ligand oxygen atoms on the same plane from the sidechain of Asp or Asn at sequence position 5 and from mainchain of carbonyl group at sequence position 7 vary from 71° to 130° (Table 2). The oxygen atom off the plane from the sidechain ligand (Asp or Asn) at sequence position 1 has O-Ca-O angles between 84° and 132°. To accommodate such great variations in O-Ca-O angle, a relatively open value with a range of 120° is used for the construction of potential sites. The O-Ca-O angles in an ideal pentagonal bipyramidal geometry for all the uni-dentate ligands on the plane and off the plane are centered at 72, 144 and 90 °, respectively. These values are close to the average values of the O-Ca-O angles in natural calcium binding proteins (Table 2). Due to the constraints for the two oxygen atoms within the same residue of Glu, the O-Ca-O angles for all bidentate ligand Glu in all three proteins are well defined with an average of 53 ± 3°. Besides the bidentate Glu and the water ligand, rotamer libraries of Asp, Asn and mainchain for 20 amino acids are used for the construction of calcium sites.

```
                         ---LOOP---
              --HELIX I--         --HELIX II-
                         1 3 5 7 9 12
                         x y z-y-x -z
EF-1 (9-39)   IAEFKEAFSLFDKDGDGTITTKELGTVMRSLGQNPT
                         *      **  *           *
EF-2 (45-75)  EAELQDMINEVDADGNGTIDFPEFLTMMARKMKDTDS
                 *        *      *  *
EF-3(82-112)  EEEIREAFRVFDKDGNGYISAAELRHVMTNLGEKLT
                        * *     **  *           *
EF-4(118-148) DEEVDEMIREADIDGDGQVNYEEFVQMMTAK
                    *     *     **  *
```

Figure 3. The locations of the residues used for the construction of the bidentate ligand in calmodulin are labeled with a star. Residues used for calcium-binding ligands are in bold. The sequence positions and relative orientations of the ligand residues are labeled above the amino acid sequences.

Identification of classic EF-hand sites in calmodulin

With the descriptions discussed above, 3772 potential sites are constructed in the calmodulin. Their locations in the primary sequence are shown in Fig. 3. When the water position is specifically placed on the top of the plane of the pentagonal geometry, the number of the constructed sites is reduced about 70%. The majority of the constructed sites are located at the four calcium binding loops while only four anchor sites are placed out of the calcium binding loops. Besides Glu at the loop position 12, which is used as the bidentate ligands for natural sites 1 to 4, loop positions 2, 8 and 9 are also used to place the bidentate Glu (anchor), where the loop position 9 in the natural calcium binding sites is often occupied by an oxygen ligand from a bridged water molecule (Linse & Forsen, 1995).

Table 1. The Ca-O lengths of the natural and relocated calcium-binding sites.

Protein	Site							
CaM			E31/Oε1	E31/Oε2	D20/Oδ	D22/Oδ	D24/Oδ	T26/O
	1	Natural	2.275	2.382	2.335	2.424	2.611	2.455
		Relocated	2.249	2.441	2.446	2.430	2.401	2.347
			E67/Oε1	E67/Oε2	D56/Oδ	D58/Oδ	N60/Oδ	T62/O
	2	Natural	2.308	2.487	2.206	2.479	2.463	2.172
		Relocated	2.375	2.469	2.407	2.367	2.388	2.340
			E104/Oε1	E104/Oε2	D93/Oδ	D95/Oδ	N97/Oδ	Y26/O
	3	Natural	2.317	2.764	2.135	2.221	2.388	2.055
		Relocated	2.156	2.452	2.425	2.180	2.423	2.079
			E140/Oε1	E140/Oε2	N129/Oδ	D131/Oδ	D133/Oδ	Q135/O
	4	Natural	2.322	2.572	2.166	2.558	2.074	2.383
		Relocated	2.256	2.185	2.986	2.554	2.412	2.104
Parv			E101/Oε1	E101/Oε2	D90/Oδ	D92/Oδ	D94/Oδ	K96/O
	2	Natural	2.486	2.505	2.246	2.415	2.439	2.294
		Relocated	2.124	2.560	2.575	2.445	2.200	2.140
CBD			E65/Oε1	E65/Oε2	D54/Oδ	D56/Oδ	D58/Oδ	E60/O
	2	Natural	2.534	2.538	2.412	2.336	2.387	2.387
		Relocated	2.461	2.482	2.301	2.309	2.459	2.246
			E27/Oε1	E27/Oε2	A14/O	E17/O	D19/O	Q22/O
	1	Natural	2.124	2.560	2.575	2.445	2.200	2.140
		Relocated	2.503	2.483	2.439	2.294	2.246	2.415

Table 2. The O-Ca-Oε angles of the natural and relocated calcium-binding sites.

Protein	Site					
			D20/Oδ	D22/Oδ	D24/Oδ	T26/O
CaM	1	Natural	99.0	73.5	149.8	129.9
		Relocated	110.9	74.3	145.1	133.0
			D56/Oδ	D58/Oδ	N60/Oδ	T62/O
	2	Natural	89.9	71.1	150.0	128.7
		Relocated	106.5	78.0	149.3	138.1
			D93/Oδ	D95/Oδ	N97/Oδ	Y26/O
	3	Natural	95.4	78.1	149.0	130.1
		Relocated	110.4	76.5	148.5	137.4
			N129/Oδ	D131/Oδ	D133/Oδ	Q135/O
	4	Natural	84.6	85.1	164.5	113.0
		Relocated	109.4	77.2	146.6	141.1
			D90/Oδ	D92/Oδ	D94/Oδ	K96/O
Parv	2	Natural	131.9	129.0	146.3	84.3
		Relocated	116.5	124.4	147.2	79.7
			D54/Oδ	D56/Oδ	D58/Oδ	E60/O
CBD	2	Natural	112.3	125.2	152.9	76.8
		Relocated	112.9	122.8	161.2	88.2
			A14/O	E17/O	D19/O	Q22/O
	1	Natural	98.7	110.0	161.8	82.1
		Relocated	86.3	125.1	150.1	71.3

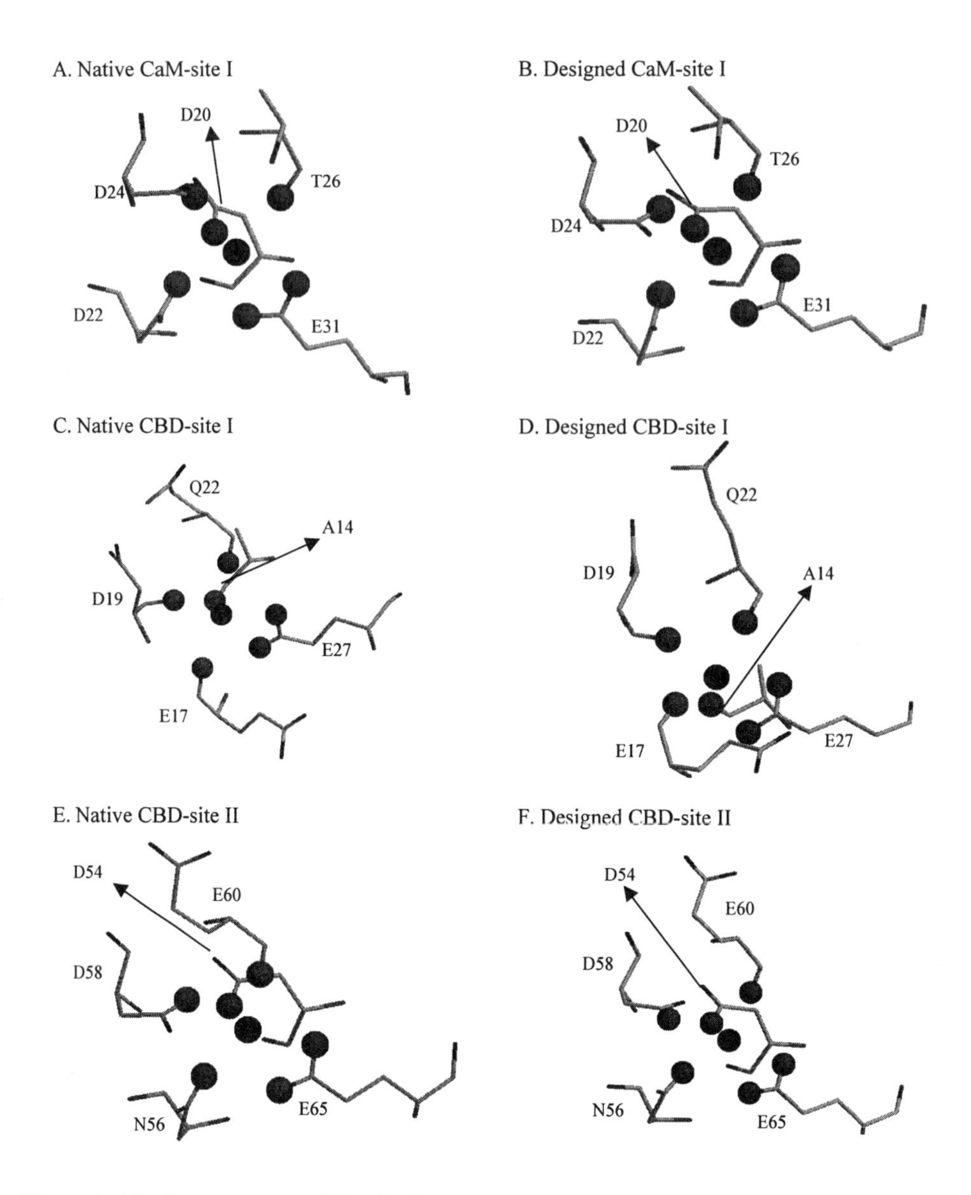

Figure 4. The local structures of the classic EF-hand calcium binding site 1 of calmodulin (A) and the relocated Site 1 (B). The local structures of the calcium binding site 1of calbindin$_{D9k}$ with a pseudo-EF-hand in the natural protein (C) and the relocated Site 1 (D). The calcium binding site 2 of calbindin$_{D9k}$ with a classic-EF-hand in the natural protein (E) and the relocated Site 2 (F).

All four natural calcium-binding sites in calmodulin are successfully identified. Their parameters for the bond length of Ca-O and angle O-Ca-O are listed in Tables 1 and 2. All the redesigned natural calcium-binding sites use the expected ligand residue and corresponding oxygen atoms. As shown in Fig. 4, the geometric parameters, such as bond lengths of Ca-O and O-Ca-O angles of the designed calcium binding sites, are very similar to that in natural calcium binding sites. Except Site 4 of calmodulin, the deviations of bond lengths of Ca-O and angles O-Ca-O are within 0.35 Å and 18 °, respectively.

Natural calcium binding sites have the lowest U(p) numbers.

The U(p) numbers for all the reconstructed calcium binding sites in calmodulin are plotted in Fig. 5. The highest number is 150.55 and the lowest one is 5.02 with a mean of 58.21. The U(p) number indicates the deviation of a site from the target site. For example, Site 3091 has the highest U(p) value of 150.55, its bond lengths vary from 1.9 to 3.7 Å and its O-Ca-O angles have 50 ° deviation from the target values. Interestingly, all natural calcium-binding sites have small U(p) numbers less than 25, suggesting small deviations from the target ideal pentagonal bipyramidal. These results are consistent with the observation from X-ray studies that the geometry of Sites 1-4 in calmodulin resemble closely the ideal bipyramidal geometry (Babu et al., 1988).

All the constructed sites with U(p) numbers close to that of natural sites share several similarities. First, they all contain the correct bidentate Glu residues from positions 31, 67, 104 and 140 in the natural calcium binding sites of calmodulin. Second, mainchain oxygen atoms from T26, T62, Y99, and Q135 in natural calcium sites (1 to 4) are always used. Third, residue numbers used for the remaining four ligand positions are the same as those in the natural calcium binding sites with either Asp replaced by Asn or Asn replaced by Asp. These calcium-binding sites are introduced mainly because both Asn and Asp are used for four positions and both residues have very similar sidechain rotamer configurations.

Table 3. Thc pseudo-energy value of relocated sites and natural sites.

Protein	Site	Natural	Relocated
CaM	1	7.84	7.47
	2	5.54	5.72
	3	11.41	11.86
	4	5.44	23.83
Parv	2	5.66	19.83
CBD	2	10.23	14.50
	1	42.67	44.56

Table 4. The relocated site 4 of calmodulin using an enlarged range of bond length.

	E140/Oε1	E140/Oε2	N129/Oδ	D131/Oδ	D133/Oδ	Q135/O
O-Ca Length *	2.400	2.412	1.248	2.878	3.080	2.593
O-Ca Length	2.312	2.311	2.369	2.389	2.336	2.347
O-Ca-O angle			80.69	82.67	153.03	128.47

* The value before minimization.

As shown in Table 3, the relocated natural calcium binding sites have small U(p) values (<25). They all have the smallest U(p) values among all the constructed sites for each protein. Therefore, the U(p) number can be used to rank the constructed potential sites according to their deviations from the target site. Our parameters used to describe the geometry of calcium binding sites allow us to accurately identify natural calcium binding sites.

To evaluate the differences between the natural sites in calmodulin, parvalbumin and calbindin$_{D9K}$ with the defined parameters of the ideal pentagonal bipyramidal geometry, the coordinates of the natural classic EF-hand sites from these three proteins are submitted into the same minimization processes. The U(p) of all the natural sites with a classic EF-hand motif are from 5-12 (Table 3). The differences of the pseudo-energy values between the natural sites and relocated sites of calmodulin site 4 and parvalbumin site 2 are noticeably greater than those of the remaining sites we examined. For example, the relocated Site 4 of calmodulin has a U(p) value of 23.08, which is a significant difference from its natural site (5.44). As shown in Fig.5, quite a few other non-natural sites have U(p) values lower than that for this related Site 4 of calmodulin. In addition, this relocated site has the largest deviation for the O-Ca length and O-Ca-O angle from that of the natural sites (Table 1 and 2). This indicates that although the parameters used are opened enough to relocate the natural ligand type, they are not enough to obtain the best configuration set. When the bond length is enlarged from 1.5-3.5 to 1.0-4.0 to research the site 4 of calmodulin, the newly relocated site 4 has the pseudo-energy of 5.26 after minimization, which almost equals that of the natural site 4. The O-Ca bond length and O-Ca-O angle are very similar to those of the ideal pentagonal bipyramidal and natural calmodulin site 4 (Table 1, 2 and 4). This newly obtained configuration of Site 4 was rejected at the Site-search step using length range of 1.5-3.5 because one of the O-Ca lengths of this relocated site (N129 Oδ-Ca) before minimization is only 1.248 Å (Table 4). It seems that the change from one configuration to another results in relatively large changes in length and angle, limited by the size of the rotamer library. Narrow parameters may reject some configurations that can be adjusted in the Refinement program.

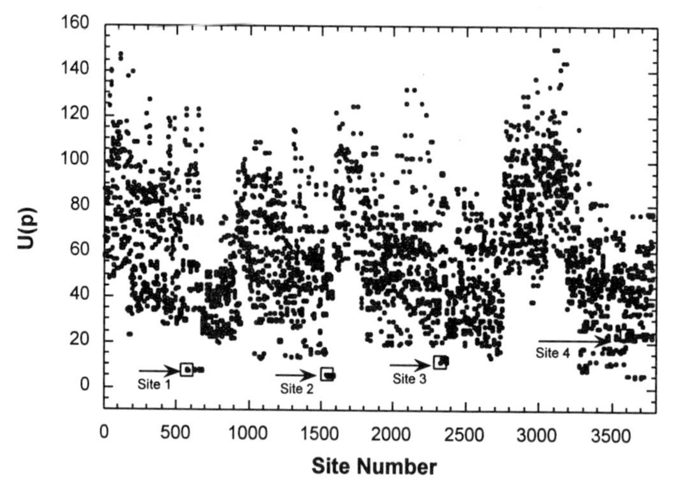

Figure 5. The deviation of the designed sites from the ideal pentagonal bipyramidal geometry (pseudo-energy U(p)) as a function the designed calcium binding sites. The natural calcium binding sites I-IV of calmodulin are shown as black dots.

The parameter set is defined by compromising between obtaining the required necessary combinations of configurations and the time used for the analysis. The parameter set with O-Ca length range of 1.5-3.5 Å and O-Ca-O angle range of 120° is suitable for the relocation of EF-hand motifs with proper ligand types and geometry in a reasonable time. To identify some specific regions of the protein with large deviation from the target site, a more opened parameter set is preferred. On the other hand, if O-Ca length range and O-Ca-O angle are specifically defined according to those of the natural site and each ligand position is occupied by its corresponding residue, the natural site can be uniquely identified in a very short time.

Identification of Pseudo EF-hand site in calbindin$_{D9k}$

We have also attempted to identify the pseudo-EF-hand site (Site 1) in calbindin$_{D9k}$. Glu is again used as an anchor residue and the water ligand is constructed by leaving one position above the plane unoccupied. Since the remaining four positions in Site 1 are from the mainchain carbonyl groups, only the coordinates of the mainchain oxygen atoms are used for the construction of the pseudo EF-hand site. With the same parameter set, only one site is constructed and all the ligand types and residue numbers for the constructed site are identical to that of the natural calcium-binding site 1. As shown in Fig. 4, the geometric parameters for both angles and both lengths for the reconstructed pseudo-EF-hand site are very similar to that of natural calcium binding site 1. Although the U(p) number for the constructed site is about 40 and larger than that for all the reconstructed natural classic EF-hand sites in calmodulin, parvalbumin and site II of calbindin$_{D9k}$, these relatively large values are generated by the relative weight factors σ_j we used for the shift of backbone atoms compared with that for the sidechain configurations in the calculation of U(p) with equation 1.

SUMMARY

Our studies for the identification of the natural calcium binding sites with both classic and pseudo EF-hand motifs in three EF-hand proteins have shown that natural calcium binding sites can be accurately relocated with Dezymer using a set of geometric descriptions of an ideal pentagonal bipyramid. The success of each constructed site can be ranked by the relative U(p) values. The searched native-like sites in three EF-hand proteins have the smallest deviation from the target geometry. Our work indicates that a useful method for searching calcium-binding sites in proteins has been established. It is possible to use established parameters to design novel calcium binding proteins.

ACKNOWLEDGEMENT

We thank Dan Adams and Anna Wilkins for the critical review of this manuscript and the helpful discussions from Drs. Don Harden and Kim Gernert, Laura Morris, Yiming Ye, Amy Carroll, Curt Coman, and the rest of the members in Dr. Jenny J. Yang's group. This work is supported by the Start-up funds, QIF, Research Initiation and Mentoring Grant for JJY at GSU.

REFERENCES

Aramini, J.M., Drakenberg, T., Hiraoki, T., Ke, Y., Nitta, K. & Vogel, H.J., 1992, Calcium-43 NMR studies of calcium-binding lysozymes and alpha-lactalbumins. *Biochemistry*. 31:6761-8

Babu, Y.S., Bugg, C.E., Cook & W.J. 1988, Structure of calmodulin refined at 2.2 angstroms resolution. *J. Mol. Biol*. 204: 191.

Bryson J.W., Betz, S.F., Lu, H.S., Suich, D.J., Zhou, H.X, O'Neil KT & DeGrado W.F., 1995, Protein design: a hierchic approach. *Science*. 270:935-41.

Clarke, N.D. & Yuan, S.M., 1995. Metal search: a computer program that helps design tetrahedral metal-binding sites. *Proteins* 23(2):256-63.

Da Silva, J.J.R.F. & Williams, R.J.P., 1991, The biological chemistry of the elements: The inorganic chemistry of life.

Falke, J.J.,Drake, S.K.,Hazard, A.L., & Peersen, O.B. 1994. Molecular tuning of ion binding to calcium

signaling proteins. *Q Rev Biophys*. 27:219-90.
Finn, B.E., Evenas, J., Drakenberg, T., Waltho, J.P., Thulin, E., & Forsen, S. 1995. Calcium-induced structural changes and domain autonomy in calmodulin. *Nat. Struct. Biol.* 2:777-83.
Glusker, J.P. 1991. Structural aspects of metal liganding to functional groups in proteins. *Adv Protein Chem*, 42:1-76.
Hellinga, H.W. 1998, The construction of metal centers in proteins by rational design. *Fold Des* 3:R1-8
Hellinga, H.W. & Richards, F.M. 1991. Construction of new ligand binding sites in proteins of known structure I. Comuter-aided modeling of sites with pre-defined geometry. *J. Mol. Biol.* 222:763-785.
Kawasaki, H. & Kretsinger, R.H., 1995, Calcium-binding proteins 1: EF-hands. *Protein Profile,* 2(4):297-490.
Klemba, M., Gardner, K.H., Marino, S., Clarke, N.D. & Regan, L., 1995, Novel metal-binding proteins by design. *Nat Struct Biol.,* 2(5):368-73.
Kretsinger, R.H., & Nockolds, C.E. 1973. Carp muscle calcium-binding protein. II. Structure determination and general description. *J Biol Chem.* 248:3313-26.
Kuboniwa, H., Tjandra, N., Grzesiek, S., Ren, H., Klee, C.B. & Bax, A. 1995. Solution structure of calcium-free calmodulin. *Nature structure biology*, 2:768-776.
Linse, S., & Forsen, S. 1995. Determinants that govern high-affinity calcium binding. *Adv Second Messenger Phosphoprotein Res*, 30:89-151.
Lu Y. & Valentine J.S., 1997, Engineering metal-binding sites in proteins. *Curr Opin Struct Biol.* 7:495-500.
Nayal, M. & Di Cera, E., 1994, Predicting Ca(2+)-binding sites in proteins. *Proc Natl Acad Sci U S A,* 91(2):817-21.
Nelson, M.R. & Chazin, W.J., 1998, Structures of EF-hand Ca(2+)-binding proteins: diversity in the organization, packing and response to Ca2+ binding. *Biometals* 11(4):297-318.
Press, W.H., Flanery, B.P., Teukolsky, S.A. & Vetterling, W.T. 1988. Numerical Recipes in C. chap.10. Cambridge University Press. Cambridge.
Regan, L. 1993, The design of metal-binding sites in proteins. *Annu Rev Biophys Biomol Struct.,* 22:257-87.
Regan, L., 1995, Protein design: novel metal-binding sites. *Trends Biochem Sci.,* 20(7):280-5.
Regan, L. & Clarke, N.D., 1990, A tetrahedral zinc(II)-binding site introduced into a designed protein. *Biochemistry,* 29(49):10878-83.
Schafer, B.W., & Heizmann, C.W. 1996, The S100 family of EF-hand calcium-binding proteins: functions and pathology. *Trends Biochem Sci.* 21:134-40.
Shi, W., Dong, J., Scott, R.A., Ksenzenko, M.Y. & Rosen, B.P., 1996, The role of arsenic-thiol interactions in metalloregulation of the ars operon. *J Biol Chem,* 271(16):9291-7.
Strynadka, N.C. & James, M.N. 1989, "Crystal structures of the helix-loop-helix calcium-binding proteins", *Annu Rev Biochem.* 58:951-98.
Svensson, L.A., Thulin, E., and Forsen, S. 1992, Proline cis-trans isomers in calbindin$_{D9k}$ observed by X-ray crystallography. *J Mol. Biol.*, 223:601.
Swain, A. L., Kretsinger R. H. & Amma, E.L. 1989, Refinement of native (calcium) and cadmium-substituted carp parvalbumin using X-Ray crystallogaphic data at 1.6 angstroms resolution. *J. Biol. Chem.* 264:16620.
Toma, S., Campagnoli, S., Margarit, I., Gianna, R., Grandi, G., Bolognesi, M., Filippis, i.D., & Fontana, A. 1991, Grafting of a calcium-binding loop of thermolysin to Bacillus Subtilis neutral protease. *Biochemistry*, 30: 97-106.
Yamashita, M.M., Wesson, L., Eisenman, G., Eisenberg, D., 1990, Where metal ions bind in proteins. *Proc Natl Acad Sci U S A,* 87(15):5648-52.

A SYNTHESIS OF FLUID DYNAMICS AND QUANTUM CHEMISTRY IN A MOMENTUM-SPACE INVESTIGATION OF MOLECULAR WIRES AND DIODES

Preston J. MacDougall [a,*] and M. Creon Levit [b]

[a] Department of Chemistry, Middle Tennessee State University
Murfreesboro, TN 37132
[b] NASA Ames Research Center, MS T27A, Moffett Field, CA 94035

INTRODUCTION

In 1959, during a famous lecture entitled "There's Plenty of Room at the Bottom",[1] Richard Feynman focused on the startling possibilities that would exist at the limit of miniaturization, that being atomically precise devices with dimensions in the nanometer range. "Molecular electronics", also refered to as "nanoelectronics", denotes the goal of shrinking electronic devices, such as diodes and transistors, as well as intergrated circuits that can perform logical operations, down to dimensions in the range of 100 nanometers.[2] The forty-year, and growing, hiatus in the development of molecular electronics can be figuratively seen as a period of waiting for the bottom-up and atomically precise construction skills of synthetic chemistry to meet the top-down reductionist aspirations of device physics. The sub-nanometer domain of nineteenth-century classical chemistry has steadily grown, and state-of-the-art supramolecular chemistry can achieve atomic precision in non-repeating molecular assemblies of the size desired for nanotechnology.[3] For molecular electronics in particular, a basic understanding of the electron transport properties of molecules themselves must also be developed. The goal of the current research is to investigate the slow (chemically valence) electron dynamics of molecules that are possible prototypes of molecular wires and diodes.[4] We refrain from basing our analysis on any of the assorted definitions of molecular orbitals. The orbital model was originally devised to explain spectroscopic properties of simple atomic systems in the gaseous phase, a phenomenon far removed from the operation of any kind of

Computational Studies, Nanotechnology, and Solution Thermodynamics of Polymer Systems
Edited by Dadmun *et al.*, Kluwer Academic/Plenum Publishers, New York, 2000

electronic device. Instead, our investigation is based on physical properties of model systems. The particular properties chosen for study are those that we expect to yield new, and general insight into the chemical factors governing electron transport at the nanoscale. This expectation arises from recent advances in the understanding of molecular structure and chemical reactivity that were obtained by analogous methods. This unexpected parallel is briefly outlined below.

A tremendous amount of information about the bonding in a molecule, and about its chemical reactivity towards other molecules, can be extracted from the molecule's electron density distribution, $\rho(\mathbf{r})$, by topological and graphical examinations of the Laplacian of this function, $\nabla^2\rho(\mathbf{r})$.[5] This method of analysis is based on an observable property of matter, the electron density, and is currently employed by both computational[6] and experimental scientists.[7] The term "electron density" is almost always assumed to refer to the probability density of electrons in the real, three-dimensional space of **r**, the position vector. Strictly speaking, $\rho(\mathbf{r})$ is the coordinate-space representation of the electronic charge density. It is simply the probability of finding *any* electron in some elemental volume, d**r**, multiplied by the total number of electrons in the system. It has the units e/a_o^3. Equally "real", and also observable, is the momentum-space representation of the electronic charge density, $\Pi(\mathbf{p})$, often called the "electron momentum density" to avoid confusion.[8] It is analogously defined as the probability of finding *any* electron in some elemental unit of momentum space, d**p**, multiplied by the total number of electrons in the system. It has the units $ea_o^3/\hbar^3$. Although they are simple to conceive, such "single-particle density distributions" are rich in information, since they are shaped by the forces and quantum mechanical symmetry requirements that are many-body in nature. For instance, even at the Hartree-Fock level the motions of all electrons with the same spin must be correlated with one another to ensure obeyance of the anti-symmetry requirement of the Pauli principle.[9] Computationally, in practice this so-called "Fermi correlation" is limited by the size of the basis set of atomic orbital-like basis functions. There is of course no such limitation on experimentally determined single-particle density distributions, the limit in this case being how many X-ray reflections, or detector coincidences, the experimentalist is able to obtain.

In keeping with the fundamental complementarity of position and momentum, a corresponding topological analysis of the Laplacian of the electron density in momentum space, $\nabla^2\Pi(\mathbf{p})$, has been shown to yield important information about the dynamics of electrons in matter.[10] In particular, the directions (with respect to the molecular framework) that slow, or valence, electrons may flow with the least effective resistance, are predicted by the regions in momentum-space near the origin where $\nabla^2\Pi(\mathbf{p})$ is most negative. In general, regions in momentum-space where $\nabla^2\Pi(\mathbf{p})$ is positive, correspond to directions within the molecule that have nonlaminar electron flow, while regions wherein $\nabla^2\Pi(\mathbf{p})$ is negative correspond to directions with laminar electron flow.[10] This local (in momentum-space) criterion for exhaustively partitioning the momentum distribution of a molecule, or crystal, into laminar and nonlaminar regions is necessarily nonlocal in coordinate-space, as demanded by the Heisenberg uncertainty principle. Thus although the $\nabla^2\Pi(\mathbf{p})$ topology and its three-dimensional isovalue surfaces presented below are relatively easy to visualize (compared to multi-dimensional wavefunctions of many-electron systems), these new entities are difficult to relate to conventional chemical concepts. The usual points of reference for chemists, nuclear positions and bond axes, are all but gone. Only the orientation of the molecule and the direction of the bond axes are directly related to fixed points in momentum-space.[11]

There has been some effort, led by Schmider, to probe the topological properties of the "fuzzy" density in phase-space for molecular systems.[12] In these studies, the electronic position and momentum are both specified, but the probabilities are uncertain, hence the "fuzziness". In this report we limit our focus to the topological properties of the well-defined $\nabla^2\Pi(\mathbf{p})$ distributions. We feel that further investigations of this function will not only foster familiarity among chemists with momentum-space itself, but more importantly such investigations will breed entirely new momentum-based concepts, such as laminar *versus* nonlaminar electronic flow. We further anticipate that continuing advances in the attainable precision of electron momentum density measurement by electron momentum spectroscopy, gamma-ray Compton scattering, and positron annihilation techniques,[13] will reinforce the knowledge gained by the analytical procedure that we discuss below.

Finally in this Introduction, we wish to reiterate a summary conclusion from earlier work,[10] but which is further supported by the current results and interim research. The extent to which the correlations among many electrons lead to local electron pairing of either the Lewis type in coordinate-space,[14] or of the Cooper type in momentum-space,[15] appears to be physically manifested as local concentration of the electronic charge density in the corresponding space. Thus charge density analysis provides a model-independent tool for relating observable information to models of electronic structure that have proven themselves paramount. The range of chemical and material behavior that have been shown to correlate with topological features of electron density distributions is not exhausted. Eberhart has recently reviewed the insight into material failure that is yielded by topological analysis of $\rho(\mathbf{r})$ in alloys.[16] We now consider analysis of the electronic charge density in momentum space of molecules that have properties that are analogous to those of electronic devices.

COMPUTATIONAL METHODS

The electronic structure package GAUSSIAN-94 was used to optimize the molecular geometries, and generate Hartree-Fock wavefunctions (with the 6-311g** basis set) for all of the molecules discussed below.[17] Calculations of perturbed wavefunctions, in the presence of a uniform, external electric field were also done (at the zero-field optimized geometries). Previous research on naphthalene and azulene compared cross-sections of $\nabla^2\Pi(\mathbf{p})$ near the origin in momentum-space (out to 1.00 au), from both Hartree-Fock and second-order Møller-Plesset calculations (also at the 6-311g** basis set).[18] Similar to studies of the effect of electron correlation on the topological properties of $\nabla^2\rho(\mathbf{r})$,[19] there were very slight numerical differences, but no topological differences between the $\nabla^2\Pi(\mathbf{p})$ distributions at these two levels of theory.

For many-electron molecules, the Hartree-Fock wavefunction that is computed by conventional electronic structure packages, such as GAUSSIAN, can be expanded from single-particle molecular orbitals, $\psi_i(\mathbf{r})$, that are themselves constructed from atom-centered gaussians that are functions of coordinate-space variables. The phase information that is contained in the molecular orbitals is necessary to define the wavefunction in momentum-space. In other words, the density in coordinate-space cannot be Fourier transformed into the density in momentum-space. Rather, within the context of molecular orbital theory, the electron density in momentum space is obtained by a Fourier-Dirac transformation of all of the $\psi_i(\mathbf{r})$'s, followed by reduction of the phase information, weighting by the orbital occupation numbers,

n_i, and finally summation over all orbitals, as shown below.[8]

$$\Pi(\mathbf{p}) = \Sigma_i\, n_i\, \left| \int d\mathbf{r}\, e^{i\mathbf{p}\cdot\mathbf{r}}\, \psi_i(\mathbf{r}) \right|^2 \quad (1)$$

This procedure is used to compute the value of the momentum density for a cubic grid of evenly-spaced (0.02 au) points in momentum-space, with $\max(p_x) = \max(p_y) = \max(p_z) =$ 1.00 au, and with the origin at the center of the cube. Electrons that are at the origin in momentum-space have no momentum. Kaijser and Smith have shown that the origin is also a center of inversion symmetry for all systems that are in a stationary state, thus only one octant of the cube is unique.[20] In our investigation, the entire cube was calculated and visual inspection confirmed a center of inversion at the origin in all cases.

In most computational studies of $\nabla^2\rho(\mathbf{r})$, the Laplacian is computed analytically since the wavefunction, and hence the density is in functional form. The Laplacian of the electron momentum density is defined analogously by,

$$\nabla^2\Pi(\mathbf{p}) = \partial^2\Pi(\mathbf{p})/\partial p_x^2 + \partial^2\Pi(\mathbf{p})/\partial p_y^2 + \partial^2\Pi(\mathbf{p})/\partial p_z^2 \quad (2)$$

However, in this work it is computed numerically since the density is not in functional form, as discussed above. The study of fluid dynamics is critical for aerodynamic design, and here too the data is collected and usually simulated, on a grid (often very complex in shape). Fluid dynamicists also routinely examine flow properties in both coordinate-space and momentum-space. The key difference being that for classical fluids the particle trajectories are known (within experimental error). Computational fluid dynamicists can also examine the momentum distribution of all particles that pass through an infinitesimally small volume of coordinate-space, and monitor the evolution of this distribution as the probe position is scanned. For quantum fluids, such as the electronic charge distribution in molecules, the Heisenberg uncertainty principle limits such knowledge to the "fuzzy" Husimi functions employed by Schmider, as discussed above.[12,21] For the momenta precision implied by the grid spacing in the current work, the uncertainty in the positions of the electrons is several times the molecular dimensions. Nevertheless, we have taken advantage of the advanced visualization techniques in the Flow Analysis Software Toolkit (FAST),[22] that was developed for classical fluids by the Numerical Aerodynamic Simulation (NAS) Systems Division at the NASA Ames Research Center. FAST numerically computes the gradient vector field for the $\Pi(\mathbf{p})$ distribution that we obtain via Eq. (1). FAST then computes the divergence of this gradient, also numerically, yielding $\nabla^2\Pi(\mathbf{p})$ with high accuracy. FAST allows numerous possibilities for data visualization, including movies. The topological transitions, as well as the overall shape of $\nabla^2\Pi(\mathbf{p})$, are sufficiently depicted by simple isovalue surfaces. In the figures presented below, $\nabla^2\Pi(\mathbf{p}) = -0.025$ au everywhere on the surface. In all the figures shown, there are no "hidden surfaces", that is $\nabla^2\Pi(\mathbf{p}) < -0.025$ au everywhere within the enclosed volumes. This particular value for the surface has no special significance. It was chosen so that no topological features would be obstructed, and it is negative so that the electron flow for all momenta *inside* the surface is *more laminar* than it is for momenta *outside* the surface.

In previous work on clusters of alkali metal atoms,[18] in addition to the very pronounced local concentration of charge at the origin in momentum-space (very laminar slow electrons), we have also observed topological features in $\nabla^2\Pi(\mathbf{p})$ outside the momentum-space volume studied in this work, which is 8.00 au^3. Sagar *et al.* have proposed a cut-off of 0.6 au for distinguishing between "slow" and "fast" momenta in atoms and monatomic ions.[23] They report fast topological features in $\nabla^2\Pi(\mathbf{p})$ that are more pronounced than the slow ones, for d-block and p-block metals, but not for nonmetals. Furthermore, in previous work on naphthalene and azulene,[18] where volumes of up to 512 au^3 were examined (with proportionately larger grid spacing), no additional regions of laminar flow were found beyond the limits of the currently computed grids. For these reasons, as well as "chemical intuition", we have focused on the slow electron dynamics in the organic systems below.

RESULTS AND DISCUSSION

Varied molecular architectures are being explored with purposes related to nanoelectronics.[2] Along with their collaborators, Tour and Reed,[4,24] have synthesized and conducted numerous experimental characterizations of molecular electronic devices that are polyphenylene-based. By controlling the size of the polymers, the substituents on the phenylene groups, and the spacers between phenylene groups, a growing family of well-defined polymers is providing exciting opportunities for nanoelectronics. The electronic characteristics of those small molecules or polymers that have either been successfully made and tested, or proposed and studied computationally, have been argued to provide a functionally complete set of logic elements.[25] In addition to the molecular "wires" that are needed for a circuit, if the circuit is to perform a simple logical function it must also possess molecular "gates". AND, OR and NOT gates form a functionally complete family of diode-based logic gates.[26] *Rectifying* diodes are selective in the direction that current can be induced to flow through them. That is, the threshold voltage above which current will flow, strongly depends on the current direction. This "bias" effectively gives them a "one-way" characteristic, regarding electron flow. Rectifying diodes can yield AND and OR gates, but not NOT gates. With applied voltage, current can also be induced to flow through *resonant tunneling* diodes (RTDs), but these do not have a directional bias. Instead, these devices "turn off" when the current-inducing voltage exceeds a second threshold value. NOT gates can be derived from combinations of both of these types of diodes.[26] Molecules **1**, **2** (X = S^-) and **3** (Y = CH_3, Z = CN) are very simple prototypes of polyphenylene-based molecular electronic device subunits that can be used in proposed diode-based digital logic circuits, such as an "adder".[27] Our purpose is *not* to investigate the practicality, or even the feasibility, of these far-off engineering applications. Rather, we have sought to employ these prototypical devices as test cases for further development of the novel theoretical procedure that was briefly reviewed above. The reader is refered to ref. 10 for further discussion of the physical basis of the analysis. As discussed above, a synthesis of results from quantum chemistry calculations with fluid dynamics concepts and visualization techniques, yielded a large amount of data.[28] We highlight key sets of data in Figs. 1 and 2. These images and other pertinent data are discussed for the individual prototypical devices below.

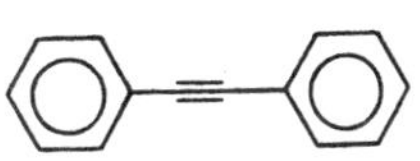

Z

X

Y

1 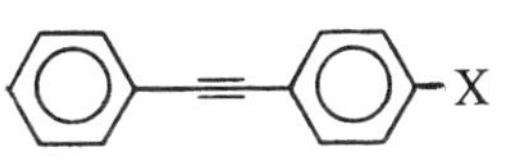 **2** 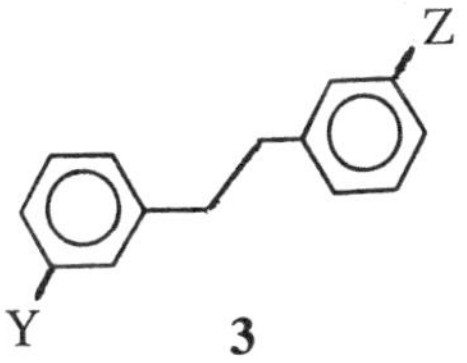 **3**

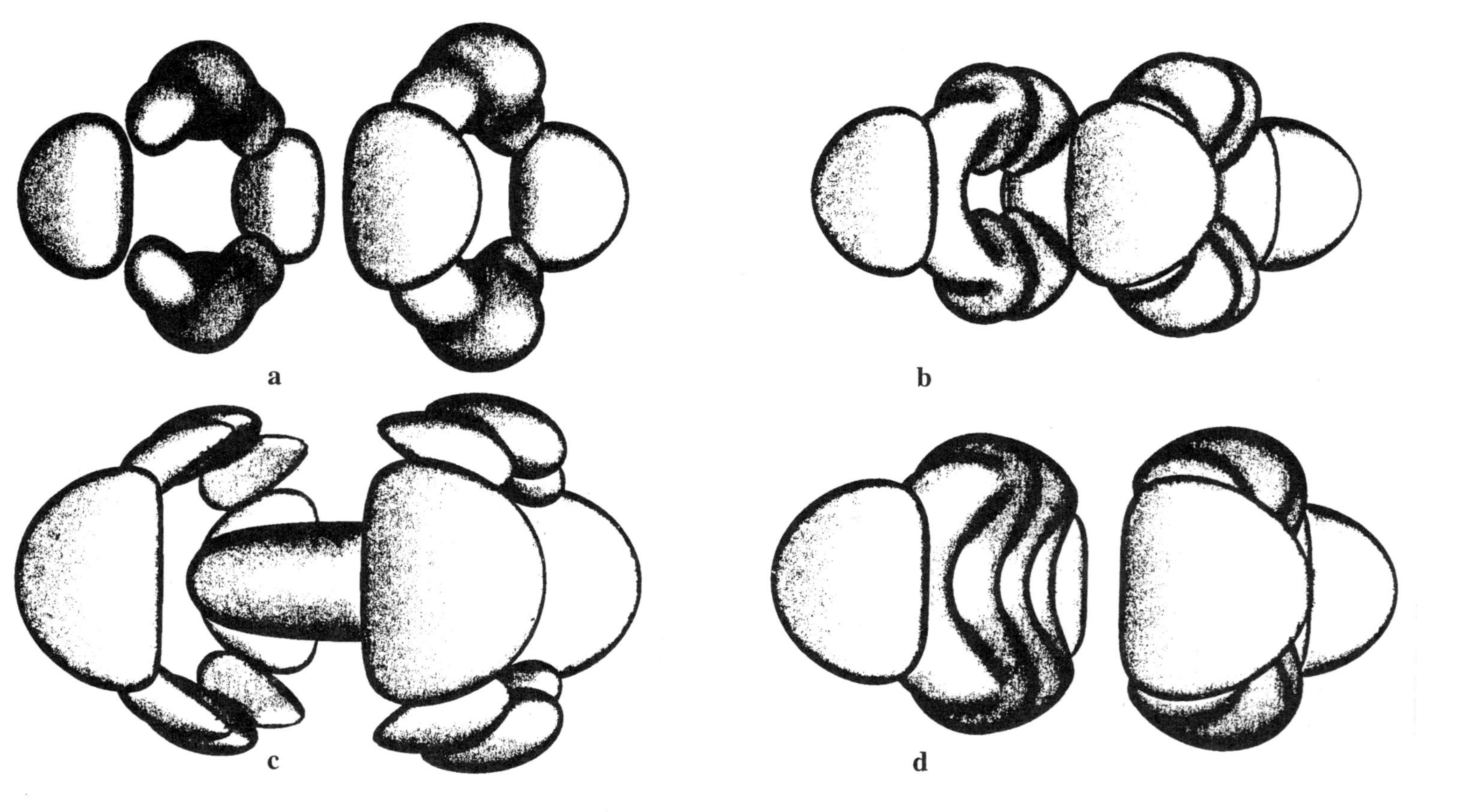

Figure 1. Isovalue surfaces of $\nabla^2\Pi(\mathbf{p})$ for (a) tolane, **1**, in the absence of an external field, (b) tolane thiolate, **2**, in the absence of an external field, (c) tolane thiolate in an external field oriented in cooperation with the electron donation of sulfur (forward bias), and (d) tolane thiolate in an external field oriented in opposition to the electron donation of sulfur (reverse bias). $\nabla^2\Pi(\mathbf{p}) = -0.025$ au everywhere on these surfaces, and $\nabla^2\Pi(\mathbf{p}) < -0.025$ au everywhere within these envelopes. The orientations of the molecules, and the momentum-space axes are as described in the "Getting Oriented" section.

Getting Oriented

Electrons with momentum parallel to the long axis in molecules **1**, **2** and **3** (which we define as the z -axis, and passes through the carbon atoms on each ring that are *para* to the bridging group) are located somewhere on the p_z -axis in momentum-space, depending on their speed. In Figs. 1 and 2, the p_z-axis is the one that appears to be coming *out of the plane* in these perspective renderings. Note that the electrons do not have to be *on* the z-axis, just possessing momentum that is parallel to it. They could be *anywhere* in coordinate-space. The x-axis in molecules **1**, **2** and **3** is defined to be perpendicular to the phenyl rings (which have a 0° dihedral angle in all of the conformations used to compute the images in Figs. 1 and 2). The p_x-axis is "horizontal" in Figs. 1 and 2, and passes through the centers of the pairs of irregularly shaped objects that appear as reflections of one another. The y-axis is thus coplanar with the phenyl rings, and in **1** and **2** it is prependicular to the triple bond. The p_y-axis is vertical in Figs. 1 and 2, and is in the plane of symmetry refered to above.

"Structure of Motion"

For tolane, **1**, we find the lowest energy conformation to be that with a 0° dihedral angle between the phenyl rings (D_{2h} symmetry), as was also reported by Seminario *at al.* using density functional methods.[29] Interestingly, we found a *very* flat potential energy surface with respect to this torsional angle. There is no significant energy difference for the 45° conformation, and R(C≡C) is constant at 1.1882 Å. When the phenyl rings are orthogonal, the molecule is only 0.42 kcal/mol less stable, and R(C≡C) decreases *very* slightly to 1.1878 Å. We find that *all* conformations of tolane, **1**, in the *absence* of an external electric field display nonlaminar slow electron dynamics (just the D_{2h} conformation is shown in Fig. 1a). This is in contrast to the very laminar behavior of the slow electrons near the origin ($\nabla^2\Pi(\mathbf{p}) << 0$) for nearly all metal atoms in the periodic table,[10] as well as diatomic and varyingly-shaped hexatomic clusters of Li and Na atoms.[18] In general, systems that would commonly be described as "metallic", have to date been found to possess laminar slow electron dynamics, while systems that would commonly be described as "nonmetallic" have been found to possess nonlaminar slow electron dynamics.[30] The key to this interpretation is that while electron-electron interactions are ubiquitous (both local and nonlocal), they are postulated to cause resistance only if the electrons are very close in momentum-space.[10] We reiterate here the important, and surprising observation that even for *monatomic* systems, the element with $\nabla^2\Pi$ closest to zero at the origin, on a per electron basis, is silicon.[10] Thus the slow electron dynamics of an isolated atom of silicon, in its ground state, presages a key property of the element in its bulk form. Furthermore, this information is extracted from just the *density* of the ground state. In Fig. 1a, regions of laminar electron flow are clearly seen at faster momenta (in the neighbor- hood of 0.5 au for *all* of the molecules studied in this work). We have not yet performed a topological analysis locating the extrema in $\nabla^2\Pi$, but we plan to do so. Although the $\nabla^2\Pi$ isolevel surface is not shown, this molecule can undergo a transition to laminar slow electron dynamics when an external electric field is turned "on". For tolane in an external electric field of 0.05 au (1 au = 1 $E_h(ea_o)^{-1} = 5.1422 \times 10^{11}$ Vm^{-1}), with a 0° dihedral angle, there is a qualitatively different topology of $\nabla^2\Pi$ near the origin. The fast concentrations seen in the absence of an external field shrink, but are still present, and a large brick-shaped concentration of charge appears at the origin. The new feature is similar in length to the cigar-shaped feature in Fig. 1c, and has a similar orientation, but it is wider in the p_z direction. In the presence of this field, the slow electrons are laminar, as in metallic systems. That is, this molecule behaves as a *semiconducting wire*, and is a prototype of what are termed "Tour wires".[4] We have not investigated the momentum-space properties of tolane at all fields up to 0.05 au, but at a field strength of 0.01 au there is no noticeable change from the surface shown in Fig. 1a.

As discussed above, for molecular electronics it is essential that some components be capable of restricting electron flow to a single direction when they are turned "on". A diode is a macroscopic electronic device that has this performance characteristic. By chemically modifying one end of a Tour wire, as shown in **2**, we introduce asymmetry in the molecule and hence in its electron dynamics. Tolane thiolate, **2**, is thus a prototype of a molecular rectifying diode. In fact, self-assembled monolayers of similar thiol-substituted Tour wires, between electrodes, have been shown to have such rectifying behavior.[32] In this case, by substituting an electron *donating* group at one end, we might intuitively expect that electron flow *from* the substituted end towards the rest of the molecule would be enhanced somehow, or "pushed" as physical organic chemists imply with arrows. Indeed, we have found that in similar electric fields that induced a transition to laminar slow electron flow in the semi-conducting wire above, we observe qualitatively different responses by the "molecular diode" **2**, depending on the direction of the applied field. Fig. 1b shows the structure of motion for the tolane thiolate ion in the absence of an applied field. The four "ear muffs" at the corners of the $p_y = 0$ plane are very similar to corresponding features in tolane. Directionally, these features correspond to diagonal motion in any plane that is parallel to the long axis *and* perpendicular to the phenyl rings, such as the π-planes of those bonds (four of them) in the rings that are parallel to the long axis. As in tolane, when a weak external electric field of 0.01 au is applied parallel to the long axis in *either* direction, there is no change in the structure of motion. When the field is oriented so that it pushes *against* the electron donation from sulfur, then the features (except the ear muffs) swell. Whereas there is virtually no change when the direction of the applied field cooperates with sulfur's electron pushing (surfaces not shown). This is in sharp contrast to what is observed in a stronger applied field of 0.025 au. Fig. 1c clearly shows a change in the structure of motion of the slow electrons when the applied field cooperates with electron donation from sulfur, while there is only further swelling of the "pancakes" for a field in the opposite direction (Fig. 1d). Thus, the slow electron dynamics of tolane thiolate display rectifying behavior at a field strength of 0.025 au, since the onset of metallic character, $\nabla^2\Pi < 0$ at the origin, has a directional bias.

The characteristic behavior of a resonant tunneling diode (RTD) is "off,on,off", as the voltage across the device is swept from low to high.[26] In macroscopic devices this is achieved by separating sources and sinks, each of which have discrete energy levels, by an insulating barrier that can be tunneled through. As the voltage is scanned, the energies of the states in the source rise, and those in the sink are lowered. Within the context of band theory, tunneling only occurs when the occupied states of the source have risen just the right amount so that they are in *resonance* with the virtual states of the sink. Further increase in the voltage disrupts this resonance, and the tunneling current shuts off. Within the context of molecular orbital theory, molecular analogues of such a device correspondingly achieve resonant tunneling via manipulation of the active, or *frontier* molecular orbital energy levels.[27] Here, the molecular

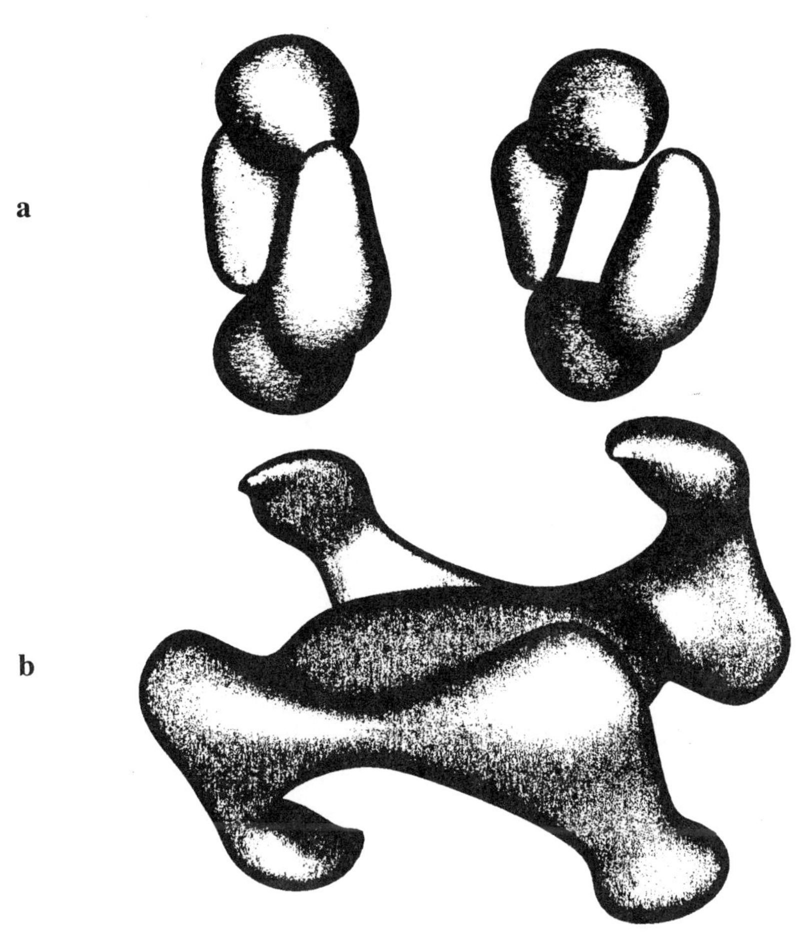

Figure 2. Isovalue surfaces of $\nabla^2\Pi(\mathbf{p})$ for a diode-like molecule, **3**, (a) field off, and (b) field on. $\nabla^2\Pi(\mathbf{p}) = -0.025$ au everywhere on these surfaces, and $\nabla^2\Pi(\mathbf{p}) < -0.025$ au everywhere within these envelopes. The orientation of the molecule, and the momentum-space axes are as described in the "Getting Oriented" section.

orbitals are described as being localized on opposite sides of an insulating barrier, such as the ethylene bridge in **3**, or other "island" that interupts the conjugation between the molecular "terminals". Molecule **3** is not a simple, or true RTD, in that it has an internal directional bias, and thus will admix rectifying behavior. True molecular RTD's have been successfully demonstrated to have "off,on,off" behavior.[24] We have not completed our investigation of the slow electron dynamics of **3**, but we report our observations for no field, and low field, since they present some interesting new features (Figs. 2a and 2b, respectively). Future work will complete the investigation of these devices. As before, we again observe a change in the structure of motion in the slow electron regime, from nonlaminar to laminar, indicating the onset of metallic behavior when the field is turned on. In this calculation the applied field is 0.05 au, in the-z direction (the applied field pushes the electrons from the cyano-bearing phenyl ring toward the methyl-substituted one). With the field on, the laminar domain near the origin is very similar in shape and orientation to those observed in the molecular wire and rectifying diode discussed above (in their "on" state). However, the laminar domains in the fast regime are qualitatively different from any other features reported so far. The "points" are particularly interesting. We also note that the internal bias of **3**, due to the electron donating (CH_3) and withdrawing (CN) groups, is in the *opposite* direction to this applied field bias. Yet, a transition to an "on" state is still observed. This might be an indication of a different physical mechanism underlying the transition in **3** as compared to **2**, however we have not investigated the effect of field reversal on **3**. Also, we have yet to determine whether or not further increase in field strength leads to another topological transition in $\nabla^2\Pi$ near the origin. Further investigation of the structure of motion in these molecular electronic devices will undoubtedly improve our understanding of the chemical control of electron transport.

SUMMARY AND CONCLUSIONS

Topological properties of the Laplacian of the electron density in momentum space were calculated from *ab initio* electronic wavefunctions for molecules that are potential components of nanoelectronic assemblies. In particular, molecules that have been proposed to function as wires and diodes were investigated with a novel computational procedure that does not, by necessity, rely on an orbital-based description of the electronic structure of the molecules, nor does it arbitrarily subtract or select data. Instead, the salient electronic flow properties were related to the locations and magnitudes of local concentrations and depletions in the total electronic charge distribution of the molecule in momentum-space. We regard the molecular electronic charge distribution as a quantum fluid that obeys uncertainty relations. Thus specific flow features cannot be ascribed to particular locations, or functional groups, within the molecule simply by refering to the atomic orbital expansion in a "chosen" molecular orbital. Nevertheless, as functional group substitutions were made, the resulting changes in the electronic flow properties of the molecule were highlighted by the Laplacian analysis, and interpreted with chemical reasoning. The important effects of external electric fields on the flow properties of the molecules were also investigated.

Specifically, we have found that an organic molecule, **1**, which in the *absence* of an external electric field displays nonlaminar slow electron dynamics, can undergo a transition to laminar slow electron dynamics when an external electric field is turned "on". That is, this molecule behaves as a semiconducting wire. We have also found that with similar electric field strengths that induced a transition to laminar slow electron flow in the semiconducting wire above, we observe qualitatively different behavior, "on" versus "off", for the "molecular diode" **2**, depending on the direction of the applied field. Thus, this molecule behaves as a rectifying diode. Interesting new features in the structure of motion are observed for **3**, which is fundamentally different in that it has a saturated, hence insulating, bridging group. Further application of this methodology seems promising.

ACKNOWLEDGMENTS

This research was conducted at the NASA Ames Research Center, supported in part by a NASA-ASEE-Stanford Summer Faculty Fellowship awarded to PJM. We thank Drs. Minhhuy Hô and Vedene H. Smith, Jr., of Queen's University, Kingston, Ontario, for sharing a Fourier transformation algorithm with us. We also thank Dr. Subhash Saini (NASA Ames) for managerial support of our initial collaboration.

REFERENCES AND NOTES

1. R. P. Feynman, *Eng. and Sci.* **23**, 22 (1960).
2. M. A. Ratner and J. Jortner, in *Molecular Electronics* , ed. J. Jortner and M. A. Ratner (Blackwell Science Ltd., London, 1997), pp. 5-72. See also references cited therein.
3. A. J. Bard, *Integrated Chemical Systems: A Chemical Approach to Nanotechnology* (John Wiley and Sons, 1994).
4. J. M. Tour, R. Wu and J. S. Schumm, *J. Am. Chem. Soc.* **113**, 7064 (1991).
5. R. F. W. Bader, P. J. MacDougall and C. D. H. Lau, *J. Am. Chem. Soc.* **106**, 1594 (1984).
6. R. F. W. Bader, *Chem. Rev.* **91**, 893 (1991), and references therein.
7. R. Destro, R. Bianchi, C. Gatti and F. Merati, *Chem. Phys. Lett.* **186**, 47 (1991). G. T. Smith, P. R. Mallinson, C. S. Frampton, L. J. Farrugia, R. D. Peacock and J. A. K. Howard, *J. Am. Chem. Soc.* **119**, 5028 (1997). R. Flaig, T. Koritsanszky, D. Zobel and P. Luger, *J. Am. Chem. Soc.* **120**, 2227 (1998). G. K. H. Madsen, B. B. Iversen, F. K. Larsen, M. Kapon, G. M. Reisner, F. H. Herbstein, *J. Am. Chem. Soc.* **120**, 10040 (1998). Y. A. Abramov, L. Brammer, W. T. Klooster and R. M. Bullock, *Inorg. Chem.* **37**, 6317 (1998).
8. R. Benesch and V. H. Smith, Jr., in *Wave Mechanics - The First Fifty Years*, ed. W. C. Price, S. S. Chissick and T. Ravensdale (Butterworth and Co., London, 1973).
9. R. McWeeny, *Rev. Mod. Phys.* **32**, 335 (1960).
10. P. J. MacDougall, *Can. J. Phys.* **69**, 1423 (1991).
11. J. Wang, B. J. Clark, H. Schmider and V. H. Smith, Jr., *Can. J. Chem.* **74**, 1187, (1996).
12. H. Schmider, *J. Chem. Phys.* **105**, 3627 (1996). H. Schmider and M. Hô, *J. Phys. Chem.* **100**, 17807 (1996).
13. M. Vos, S. A. Canney, I. E. McCarthy, S. Utteridge, M. T. Michalewicz and E. Weigold, *Phys. Rev. B* **56**, 1309 (1997). J. Nakamura, T. Takeda, K. Asai, N. Yamada, Y. Tanaka, N. Sakai, M. Ito, A. Koizumi and H. Kawata *J. Phys. Soc. Japan* **64**, 1385 (1995). S. Ishibashi, A.A. Manuel, L. Hoffmann and K. Bechgaard, *Phys. Rev. B* **55**, 2048 (1997).
14. G. N. Lewis, *J. Am. Chem. Soc.* **38**, 762 (1916).
15. L. N. Cooper, *Phys. Rev.* **104**, 1189 (1956).
16. M. E. Eberhart, *Scientific American* , **281**, 66 (1999). M. E. Eberhart and A. F. Giamei *Materials Sci. Eng.* **A248**, 287 (1998).
17. Gaussian 94 (Revision D.3), M. J. Frisch, G. W. Trucks, H. B. Schlegel, P. M. W. Gill, B. G. Johnson, M. A. Robb, J. R. Cheeseman, T. A. Keith, G. A. Petersson, J. A. Montgomery, K. Raghavachari, M. A. Al-Laham, V. G. Zakrzewski, J. V. Ortiz, J. B. Foresman, C. Y. Peng, P. Y. Ayala, M. W. Wong, J. L. Andres, E. S. Replogle, R. Gomperts, R. L. Martin, D. J. Fox, J. S. Binkley, D. J. Defrees, J. Baker, J. P. Stewart, M. Head-Gordon, C. Gonzales, and J. A. Pople, Gaussian, Inc., Pittsburgh, PA (1995).
18. P. J. MacDougall, "On the gradient path towards design tools for nanotechnology that are based on the electron density" New Technology Seminar, NASA Ames Research Center (June 9, 1997), available on video from http://www.nas.nasa.gov/Services/DocCenter/Videos/ .
19. C. Gatti, P. J. MacDougall and R. F. W. Bader, *J. Chem. Phys.* **88**, 3792 (1988).
20. P. Kaisjer and V. H. Smith, Jr., in *Quantum Science, Methods and Structure*, ed. J.-L. Calais (Plenum Press, New York, 1976).
21. K. Husimi, *Proc. Phys. Math. Soc. Jpn* **22**, 264 (1940). J. E. Harriman, *J. Chem. Phys.* **88**, 6399 (1988). R. C. Morrison and R. G. Parr, *Int. J. Quantum Chem.* **39**, 823 (1991).

22. FAST was created by V. Watson, F. Merritt, T. Plessel, R. K. McCabe, K. Castegnera, T. Sandstrom, J. West, R. Baronia, D. Schmitz, P. Kelaita, J. Semans and G. Bancroft, at the NASA Ames Research Center, NAS Division. Information about FAST is at http://www.nas.nasa.gov/FAST/fast.html .
23. R. P. Sagar, A. C. T. Ku, V. H. Smith, Jr. and A. M. Simas, *J. Chem. Phys.* **90**, 6520 (1989).
24. M. A. Reed, *Proc. IEEE* **87**, 652 (1999).
25. D. Goldhaber-Gordon, M. S. Montemerlo, J. C. Love, G. J. Opiteck, and J. C. Ellenbogen, "Overview of Nanoelectronic Devices", *Proc. IEEE* **85**, 521 (1997).
26. R. C. Jaeger, *Microelectronic Circuit Design* (McGraw-Hill, New York, 1997).
27. J. C. Ellenbogen and J. C. Love, "Architectures for molecular electronic computers: 1. Logic structures and an adder built from molecular electronic diodes", MITRE Technical Report No. 98W0000183, The MITRE Corporation, McLean, VA, July 1999.
28. Archival data from GAUSSIAN calculations (atomic coordinates, Hartree-Fock energies, dipole moments, etc.) are available upon request.
29. J. M. Seminario, A. G. Zacarias and J. M. Tour, *J. Am. Chem. Soc.* **120**, 3970 (1998).
30. Known exceptions to this general case are instructive. The ground states of Pd and Pt atoms (taken to be ^1S in ref. 10) have nonlaminar electron dynamics at the origin in momentum-space. These states differ from all other metal atoms in that the outermost electron shell has an "inert" 18-electron configuration. The only other two metals found to posses nonlaminar dynamics at the origin were Bi and Po. This may be related to the use of nonrelativistic ground state wavefunctions. Recently, Essén has elegantly argued that under special circumstances, magnetic effects of relativistic origin can lead to significant macroscopic consequences, particularly in metals and plasmas.[31]
31. H. Essén, *Phys. Rev. E* **53**, 5228 (1996). H. Essén, *J. Phys. A: Math. Gen.* **32**, 2297 (1999).
32. C. Zhou, M. R. Deshpande, M. A. Reed and J. M. Tour, *Appl. Phys. Lett.* **71**, 611 (1997).

CLASSICAL AND QUANTUM MOLECULAR SIMULATIONS IN NANOTECHNOLOGY APPLICATIONS

Robert E. Tuzun

Computational Science Program
SUNY Brockport, Brockport, NY 14420

ABSTRACT

Increases in computer power and improvements in algorithms have greatly extended the range of applicability of classical molecular simulation methods. In addition, the recent development of Internal Coordinate Quantum Monte Carlo (ICQMC) has allowed the direct comparison of classical simulations and quantum mechanical results for some systems. In particular, it has provided new insights into the zero point energy problem in many body systems. Classical studies of non-linear dynamics and chaos will be compared to ICQMC results for several systems of interest to nanotechnology applications. The ramifications of these studies for nanotechnology applications will be discussed.

INTRODUCTION

Because many details of the dynamics and structure of chemical systems cannot be directly observed, molecular simulation methods such as molecular dynamics (MD) [1-3], molecular mechanics (MM) [4], and classical and quantum Monte Carlo [5,6] are extremely valuable tools for making sense of experimental results. In the context of nanotechnology, molecular simulation is crucial for studying the feasibility of proposed directions of research and development [7]. With the rapid improvement in computing power and algorithms, the capabilities and range of applicability of molecular simulation have dramatically increased over the past decade.

This article describes work performed at Oak Ridge National Laboratory on the development of molecular simulation methods. Our group has optimized molecular simulation methods particularly suitable for systems with highly interconnected bond networks. One area of improvement is in the calculation of forces and second derivatives

Computational Studies, Nanotechnology, and Solution Thermodynamics of Polymer Systems
Edited by Dadmun *et al.*, Kluwer Academic/Plenum Publishers, New York, 2000

common to many types of simulation [8-11]. Another is a method for automatically keeping track of bend, torsion, and other interactions in any bond network [12]. A more recent development is the development of internal coordinate quantum Monte Carlo (ICQMC) [13], a method for the calculation of vibrational energy levels and wavefunctions of systems with many atoms.

This work has begun to yield fruit in the study of quantum- classical correspondence in many body systems. Although considerable work has gone into this area for few body systems, comparatively little work has been done for many body systems. Results so far indicate that classical MD simulations can grossly overestimate vibrational motion in many body systems due to the leakage of zero point energy into high amplitude vibrational modes. Although this problem is mitigated by geometric constraints or by approximations that reduce the number of degrees of freedom (such as collapsed atoms), comparison of classical and quantum simulations underscore the need for the determination of the limits of classical simulation. Although such studies have led to two schemes for correcting the zero point energy problem in few body systems, there are many challenges in extending these methods to larger systems.

It should be noted that this problem is much less of an issue in, for example, classical Monte Carlo calculations, where successive iterations do not correspond to a progression of time steps. Similar considerations apply to molecular mechanics calculations. The consequences for the MD simulation of liquids, which are unbounded, are also much less severe than for chemically bound systems.

The next section describes improvements in general methods for calculating potential energy first and second derivatives, which are applicable to almost all molecular simulations. After this, the recently developed ICQMC method is described, followed by a discussion of classical and quantum results in the simulation of polymer chains and of carbon nanotubes.

CLASSICAL SIMULATION METHODS

The potential energy surfaces of biological, polymer, and other systems of interest to nanotechnology applications are usually written as sums of interactions between chemically bonded atoms (stretch, bend, torsion, improper torsion, and so on) and non-bonded interactions. MD and other molecular simulation methods require first and sometimes second derivatives of potential energy terms. A potential energy term $V(\phi)$, where ϕ is an internal coordinate, has the following first and second derivatives:

$$\frac{\partial V(\phi)}{\partial q_i} = \frac{\partial V}{\partial \phi}\frac{\partial \phi}{\partial q_i} \tag{1}$$

$$\frac{\partial^2 V(\phi)}{\partial q_i \partial q'_j} = \frac{\partial V}{\partial \phi}\frac{\partial^2 \phi}{\partial q_i \partial q'_j} + \frac{\partial^2 V}{\partial \phi}\frac{\partial \phi}{\partial q_i}\frac{\partial \phi}{\partial q'_j} \tag{2}$$

Although there are usually many more non-bonded than bonded interactions, because of the more complicated functional forms the effort for the bonded part can still be an appreciable fraction of the total. Recognizing this, many research groups have put considerable effort into speeding up the calculation of internal coordinates and derivatives for stretch and other chemically bonded interactions (see, for example, [14-18]).

Our group has found it most advantageous computationally to take advantage of the highly connected nature of the bond networks for the systems we simulate [8-11]. For example, in a diamondoid bond network one bond stretch can be part of as many as six different bend interactions and 24 different torsion interactions. It therefore makes sense to use two- and three-body internal coordinates and derivatives as intermediates for those for three- and four-body interactions. A bond distance and its derivatives can be used as intermediates for bond angles and their derivatives, thus avoiding duplicate calculations, most especially square roots. For example,

$$\cos\theta_{123} = \frac{\partial r_{12}}{\partial x_1}\frac{\partial r_{23}}{\partial x_3} + \frac{\partial r_{12}}{\partial y_1}\frac{\partial r_{23}}{\partial y_3} + \frac{\partial r_{12}}{\partial z_1}\frac{\partial r_{23}}{\partial z_3} \tag{3}$$

$$\frac{\partial \cos\theta_{123}}{\partial q_1} = -\frac{1}{r_{12}}\left(\frac{\partial r_{12}}{\partial q_1}\cos\theta_{123} + \frac{\partial r_{23}}{\partial q_2}\right) \tag{4}$$

where q is x, y, or z. Savings are even more dramatic for torsion and other four-body interaction types.

Translational invariance and symmetry of second derivatives are commonly used to speed up simulation calculations. In addition, simple expressions for gradients of internal coordinates we derived for ICQMC calculations can also be used to speed up MM and other calculations. For example, once the second x and y derivatives of a bond angle are calculated, we obtain the z derivative by

$$\nabla_1^2 \cos\theta_{123} = -\frac{2\cos\theta_{123}}{r_{12}^2} \tag{5}$$

$$\frac{\partial^2 \cos\theta_{123}}{\partial z^2} = \nabla_1^2 \cos\theta_{123} - \frac{\partial^2 \cos\theta_{123}}{\partial x^2} - \frac{\partial^2 \cos\theta_{123}}{\partial y^2} \tag{6}$$

One difficulty our group has addressed is the bookkeeping required for keeping track of internal coordinates and their derivatives in different types of bond networks. Previously it was necessary to account for the type of bond network in order to optimize simulation codes. For example, it makes sense to think of carbon nanotubes in terms of rings, polymer chains in terms of monomer units, and so on. However, codes constructed in this manner are not easily portable to different applications. We have found a way to automatically construct tables of stretch, bend, torsion, and other interactions for any bond network beginning with a connection table [12]. The tables are split according to the numerical order of the atom numbers (for example, 123 vs. 213 vs. 312 for bend interactions). Once this is done, our formulas for internal coordinates and derivatives can be implemented with no loss of efficiency. In addition, the code can be used for different applications with little or no modification. Although it is still necessary to account for the functional form of the potential energy surface, the general bond network method can aid in the classification of interaction types (for example, according to atom types in the most general codes).

INTERNAL COORDINATE QUANTUM MONTE CARLO

Very few quantum mechanical problems can be solved exactly; problems in this category include certain types of single oscillators or systems of uncoupled oscillators. To solve the Schrodinger equation for systems of chemical interest (i.e., many-body systems with strongly coupled potential energy surfaces) requires approximate methods. Although many of these methods yield accurate results for small molecules, the computational effort scales so steeply with size as to become intractable for polymer, biological, and other highly interesting systems.

Perhaps the most promising method for solving the Schodinger equation for many body systems is quantum Monte Carlo (QMC). This essentially consists of rewriting the Schrodinger equation as a diffusion equation in imaginary time. This approach has proven very successful in electronic structure and liquid structure problems. Only recently, through the development of internal coordinate QMC (ICQMC), has it become possible to apply this method to polymer and other many-atom chemical systems. The Schrodinger equation can be written

$$\frac{\partial \Psi}{\partial \tau} = \left\{ \sum_i D_i \nabla_i^2 - V \right\} \Psi \tag{7}$$

which is a diffusion equation with diffusion coefficients

$$D_i = \frac{\hbar^2}{2m_i} \tag{8}$$

plus a first order rate term. This can be solved by generating a set of random walkers (also called psi particles) that are diffused according to the diffusion constants and created or annihilated according to the first order rate constant. It can be formally proven that the solution converges to the ground state wavefunction. By using symmetry and orthogonalization methods it is possible to modify this approach to get excited states.

This equation is most suitable for few-body systems. It should be noted that creation/annihilation of random walkers depends only on the total potential energy. In many-body systems this becomes insensitive to individual variations in bond lengths, angles, etc. This means that a larger problem becomes less controllable. The use of importance sampling [19] introduces a level of control that allowed larger systems to be treated. The essential idea of importance sampling is to introduce a guiding function (also called a trial function) ψ_o and to define ψ by $\Psi = \psi/\psi_o$. Introducing this trial function and defining $\tau = it/\hbar$ yields

$$\frac{\partial \psi}{\partial \tau} = \sum_i D_i \left[\nabla_i^2 \psi - \nabla_i \cdot (\psi \nabla_i \ln \psi_o) \right] - \left(\frac{\hat{H} \psi_o}{\psi_o} \right) \psi \tag{9}$$

The first and last terms on the right hand side are diffusion and first order terms. The middle term corresponds to what is called drift or quantum force. In every iteration, this term forces each random walker into regions of higher trial wavefunction density, in much the same way that classical forces move atoms in MD simulations. This adds a degree of controllability that allows systems with at least several hundred particles to be treated: the more closely the trial wavefunction resembles the true wavefunction, the faster the calculation will converge.

Since the potential energy surfaces of chemical systems are usually written in terms of internal coordinates such as bend and torsion, it makes sense to write the trial function in terms of internal coordinates also. On physical grounds, we write the trial function as a product

$$\psi_o = \prod_i \chi_i(\phi_i) \tag{10}$$

where ϕ_i are the internal coordinates and the ϕ's are Gaussians. The calculation requires gradients and Laplacians of the internal coordinates. Our expressions for Laplacians are especially important here: because they can be written in terms only of internal coordinates, it is unnecessary to compute and add individual second derivatives. These expressions make it practical to perform ICQMC on modest workstations and high-end PC's.

COMPARISON OF CLASSICAL AND QUANTUM RESULTS

Until recently, QMC had not been applied to many body chemical systems and direct comparison of classical and quantum simulations was not possible. Recently, it was found that simple diffusion QMC (no importance sampling) performed well on polyethylene chains with 50 or fewer atoms [20]. Beyond this number, the method failed due to the controllability issue discussed above.

ICQMC was first applied to united atom model polyethylene chains with 100 monomer units; typical results are shown in Fig. 1 [21]. The ground state energy is slightly below the normal modes result. In addition, as expected for the ground state, wavefunctions have a nearly Gaussian shape. As shown by the end to end distance distribution, the ground state is quite rigid; the width of the end to end distance distribution is of the same order as those of the individual bond lengths and is centered near the equilibrium value. It should be emphasized that the trial wavefunction constrains only internal coordinates; no restrictions are placed on the end to end distance, unlike Cartesian trial functions.

So far, we have calculated ground states for polyethylene chains with up to 400 atoms. For 100 or fewer atoms, it takes several thousand time steps to converge. As the system gets larger, the number of iterations required for convergence increases sharply. For 400 atom chains, it took upwards of 100,000 iterations to converge. We are attempting to address this issue by improving the quality of the trial functions (by including local coupling effects, for instance).

These results are in sharp contrast with classical MD simulation results. Fig. 2 shows end to end distance profile for a polyethylene chain at (a) 100% and (b) 25% of the zero point energy. At these energies, the polymer chain undergoes either significant coiling or large amplitude motion. This occurs because energy which should be locked in placed quantum mechanically flows freely in classical simulations, subject only to the conservation of energy. Any vibrational energy that enters low-frequency, high-amplitude modes may cause unrealistically large amplitude motion. This phenomenon is known as the zero point energy problem or adiabatic leak. The significance of these results is underscored by a previous study which reported high-dimensional chaos in MD simulations of polyethylene chains even at temperatures below 2K [22]. In few-body systems, it is possible to construct classical trajectories that are quasi-periodic and that mimic the behavior of quantum states; no such trajectories have yet been found for polyethylene chains.

Admittedly, these results pertain to a worst case scenario: a one-dimensional bond network with no external constraints. Similar comparisons have been made for carbon nanotubes [23], which are much more highly connected. In this case, the agreement is much better in that the nanotube keeps its shape in the MD simulations; however, the magnitude of longitudinal motion and ring breathing is still far in excess of the quantum result.

Essentially, the problem of adiabatic leak is mitigated by the presence of constraints, whether in the form of steric hindrance, external forces, or in reduction of dimensionality (for example, simulations in torsion space). In simulations of proposed nanomachines, for example, rigid body dynamics is sometimes applied to components or substructures in order to make the entire nanostructure more rigid. All of this suggests that the prediction of failure modes in some nanomachine designs *due strictly to vibrational motion* is overly pessimistic. However, this does not by any means minimize concerns such as the difficulty of building desired nanostructures, chemical reactivity, and so on.

It should be noted that we are comparing the results of classical, constant energy simulations with quantum results. In many kinds of simulations, zero point energy is far less of a concern. For example, classical Monte Carlo calculations are used, among other applications, to calculate equilibrium structures of polymer micelles and other formations; in this case, the iterations do not correspond to a progression of time steps.

Several attempts have been made to correct the zero point energy problem; however, these have been applied so far only to few-dimensional systems such as Henon-Heiles and an idealized model of water [24-27]. In several of these schemes, energies of single vibrational modes are prevented from falling below their zero point values. This means not entering an elliptical region of (p, x) space; the correction schemes differ according to how this is done. For example, Bowman and co-workers [23,24] reverse the momentum of a mode when the ellipse is reached; McCormack and Lim [25,26] slide around the ellipse and move away on the "other side". To adapt these or other schemes to many-body systems will very likely require accounting for specific features of the problem.

CONCLUSIONS

Recent years have seen significant improvements to algorithms for molecular simulation. Many of these improvements are in the calculation of potential energy first and second derivatives, which are generally applicable to all molecular simulations. Computational schemes specially adapted for highly connected bond networks are discussed here.

The recent development of internal coordinate quantum Monte Carlo has made it possible to directly compare classical and quantum calculations for many body systems. Classical molecular dynamics simulations of many body systems may sometimes overestimate vibrational motion due to the leakage of zero point energy. The problem appears to become less severe for more highly connected bond networks and more highly constrained systems. This suggests that current designs of some nanomachine components may be more workable than MD simulations suggest. Further study of classical-quantum correspondence in many body systems is necessary to resolve these concerns.

References:

[1] M. L. Klein, *Ann. Rev. Phys. Chem.* **36**, 525 (1985).

[2] W. G. Hoover, *Ann. Rev. Phys. Chem.* **34**, 103 (1986).

[3] M. P. Allen and D. J. Tildesley, *Computer Simulation of Liquids*, Clrendon Press, Oxford, 1987.

[4] U. Burkert and N. L. Allinger, *Molecular Mechanics*, American Chemical Society, Washington, DC, 1982.

[5] M. Vacatello, *Macromol. Theory Simul.* **6**, 613 (1997) and references therein.

[6] B. L. Hammond, W. A. Lester, Jr., and P. J. Reynolds, *Monte Carlo Methods in Ab Initio Quantum Chemistry*, World Scientific, Singapore, 1994.

[7] K. E. Drexler, *Nanosystems: Molecular Machinery, Manufacturing, and Computation*, John Wiley, New York, 1992.

[8] R. E. Tuzun, D. W. Noid, and B. G. Sumpter, *Macromol. Theory Simul.* **4**, 909 (1995).

[9] R. E. Tuzun, D. W. Noid, and B. G. Sumpter, *Macromol. Theory Simul.* **5**, 771 (1996).

[10] R. E. Tuzun, D. W. Noid, and B. G. Sumpter, *J. Comput. Chem.* **18**, 1805 (1997).
[11] R. E. Tuzun, D. W. Noid, and B. G. Sumpter, *J. Comput. Chem.* (in press).
[12] R. E. Tuzun, D. W. Noid, and B. G. Sumpter, *J. Comput. Chem.* **18**, 1513 (1997).
[13] R. E. Tuzun, D. W. Noid, and B. G. Sumpter, *J. Chem. Phys.* **105**, 5494 (1996).
[14] K. J. Miller, R. J. Hinde, and J. Anderson, *J. Comput. Chem.* **10**, 63 (1989).
[15] H. Bekker, H. J. C. Berendsen, and W. F. vanGunsteren, *J. Comput. Chem.* **16**, 527 (1995).
[16] B. Jung, *Macromol. Theory Simul.* **2**, 673 (1993).
[19] D. Ceperly, *J. Comput. Phys.* **51**, 404 (1983).
[20] S. N. Kreitmeier, D. W. Noid, and B. G. Sumpter, *Macromol. Theory Simul.* **5**, 365 (1997).
[21] R. E. Tuzun, D. W. Noid, and B. G. Sumpter, *Macromol. Theory Simul.* **5**, 203 (1998).
[22] D. E. Newman, C. Watts, B. G. Sumpter, and D. W. Noid, *Macromol. Theory Simul.* **6**, 577 (1997).
[23] D. W. Noid, R. E. Tuzun, and B. G. Sumpter, *Nanotechnology* **8**, 119 (1997).
[24] J. M. Bowman, B. Gazdy, and Q. Sun, *J. Chem. Phys.* **91**, 2859 (1989).
[25] W. H. Miller, W. L. Hase, and C. L. Darling, *J. Chem. Phys.* **91**, 2863 (1989).
[26] K. F. Lim and D. A. McCormack, *J. Chem. Phys.* **102**, 1705 (1995).
[27] D. A. McCormack and K. F. Lim, *J. Chem. Phys.* **106**, 572 (1997).

COMPUTATIONAL DESIGN AND ANALYSIS OF NANOSCALE LOGIC CIRCUIT MOLECULES

Kimberley K. Taylor, Dan A. Buzatu, and Jerry A. Darsey†

Department of Chemistry
University of Arkansas at Little Rock
Little Rock, AR 72204

INTRODUCTION

There is a practical limit to how much can be packed onto a silicon chip. Present silicon chip technology is quickly closing in on this limit. "Physics may soon impose barriers that could slow the chips industry's blazing progress to a crawl."[1] There is a need for the development of other technologies that will allow further miniaturization of electronics. One way this can be achieved is by using molecules to construct electronic circuits. Recently there have been numerous examples in the literature of nanoscale molecules with electronic properties.[2-8] The work that is presented here proposes using molecular components, which have electronic properties, to construct a nanoscale molecular electronic logic circuit.[9] It is based on the concept of the molecular generator, which is one of the basic building blocks of this molecular circuit. The molecular nanomotor, from which we designed our molecular generator molecules, was first

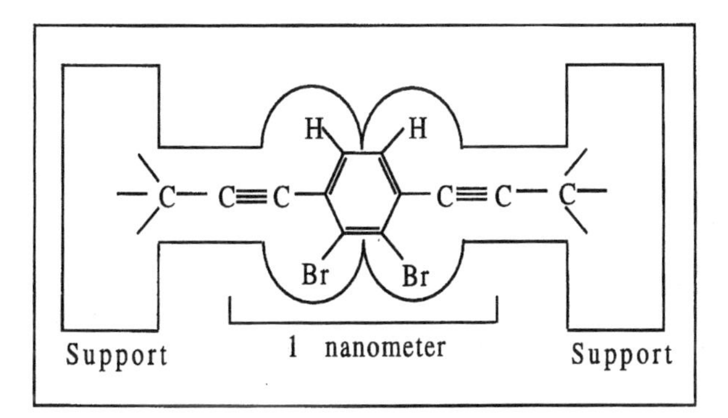

Figure 1. Schematic of a nanogenerator molecule.

† Author to whom correspondence should be addressed.

proposed by K.E. Drexler.[10] Figure 1 shows the schematic of a nanogenerator molecule. The idea behind the nanogenerator molecule is a rotating benzene ring attached to conjugated polymers through σ bonds. If the molecule is placed in a magnetic field perpendicular to the plane of the molecule, the rotating ring should produce an electrical potential that will induce a current in the overall molecule. Furthermore, different substituents such as halogens can be attached to each ring to alter the ring's resonant frequency. Circularly polarized light of different frequencies could then be used to stimulate rotation of the benzene ring on either one side of the molecule or the other This could be used to control the direction of the current in the molecular circuit. Figure 2 shows the schematic diagram of a potential logic circuit molecule, which incorporates the molecular generator molecule in Figure 1.

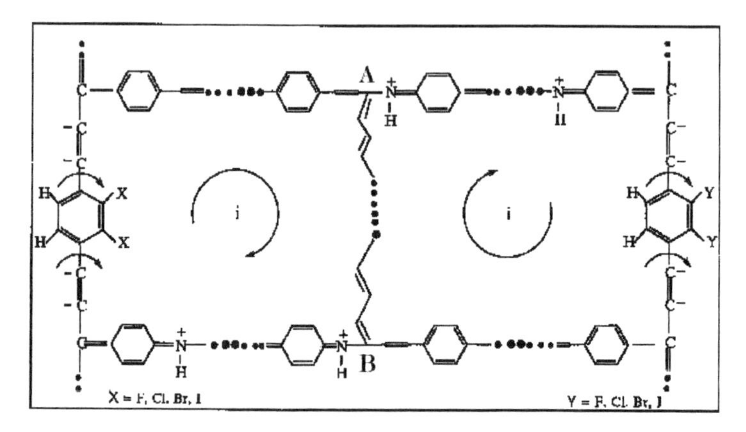

Figure 2. Nanoscale logic circuit molecule.

The logic circuit molecule is composed of two disubstituted phenyl rings connected to segments of polyaniline, polyphenylene vinylene, and polyacetylene polymer molecules. There is nothing unique about the conductive polymers chosen, and other good conductive polymer could be used. If each benzene ring is substituted with a different halogen at the X and Y positions (i.e. X=Cl and Y=F), then different frequencies of polarized light could be used to separately excite each of the molecular generators.

The frequencies of light required to excite rotation of the substituted benzene rings reside in the microwave frequency range. This will be demonstrated a little later. As an example, consider that microwave radiation of the particular frequency is used that only excites the substituted ring on the left side of the schematic in Figure 2. Let us also assume that there is a magnetic field coming perpendicular to and through the plane of the page directly at us. Rotation of the ring in the magnetic field in the indicated direction will cause a force to be exerted on the "free" electrons, to cause a small current to flow in the clockwise direction, as indicated by the arrow in Figure 2. This will cause electrons to flow to point A and then down the polyacetylene bridge to point B. It is important to note here that polyacetylene is used as a resistor in the circuit. This is based on the fact that the conductivity of bulk, undoped, trans-polyacetylene polymer is approximately 10^{-5} S/cm. The conductivity of the acid doped polyaniline polymer is approximately 10^{0} to 10^{1} S/cm.[11] The resistance encountered by the electrons moving from A to B will cause more electrons to build up at point A, and less electrons at point B. Thus point A will be more negative, and point B will be more positive. Similarly if only the ring on the right side of the molecule in Figure 2 is excited, the rotation of the ring will cause a clockwise current in this side of the molecule. This in turn will cause current to flow from point B to point A across the polyacetylene bridge, which has the effect of making B more negative than A. This characteristic (the reversibility of charge on A and B according to frequency) will allow the circuit to function as a "yes" or "no" switch, or in other words as a basic logic component.

CALCULATIONS

Electrical Properties

We treated each benzene ring in the generator portions of the molecule in the approximation of a single wire loop, with the benzene ring's conjugated π system acting similar to a wire. If the benzene ring was placed in a magnetic field perpendicular to its plane as shown in Figure 3 then its rotation in this magnetic field would give rise to the following force,

$$E_0 = NBA\,\omega \qquad (1)$$

where N = number of turns in the loop, B is the magnitude of the magnetic field, A is the area of the loop, and ω is the frequency of rotation for the loop. A benzene molecule was constructed using Hyperchem 5.01[12] and then saved to a z-matrix file. This file was then imported into the Gaussian94[13] program for geometry optimization. The molecule was then geometry optimized under Gaussian94 using the 6-31g* basis set, and the Berni optimization algorithm. The carbon bond lengths in the benzene ring where determined to be 1.385 Å. From this distance the area A in the formula for the benzene ring was calculated to be 9.967×10^{-20} m^2. A value of $N = 1$ was used for the number of loops, and $\omega = 1.8 \times 10^{12}$ sec^{-1} was the oscillation frequency used for the dichloro substituted benzene ring pictured in Figure 3. This frequency was obtained from dynamics simulations, which will be discussed in the next section. If the dichlorobenzene ring is placed in a 10 Tesla magnetic field, the rotation of the ring will produce a potential (E_0) of 1.794 μV peak, 1.268 μV rms. Similarly when a value of $\omega = 3.6 \times 10^{11}$ sec^{-1} for the difluoro substituted benzene ring in a 10 T magnetic field will produce a potential E_0 = 0.358 μV peak, and 0.253μV rms.

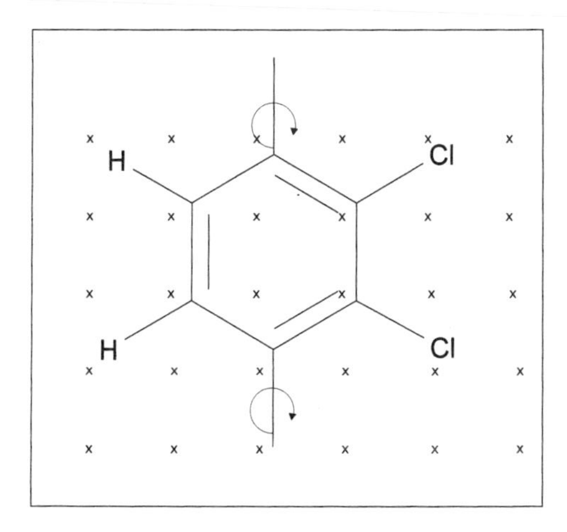

Figure 3. Dichloro benzene ring in a magnetic field coming out of the plane of the paper.

Molecular Modeling and Dynamics

The molecule was constructed and modeled using *Sybyl*[14] on a Silicon Graphics Workstation. Figure 4 illustrates a wire diagram of the molecule without the hydrogen atoms after it was energy minimized using the interface to MOPAC, and the AM1 parameter set under Sybyl.

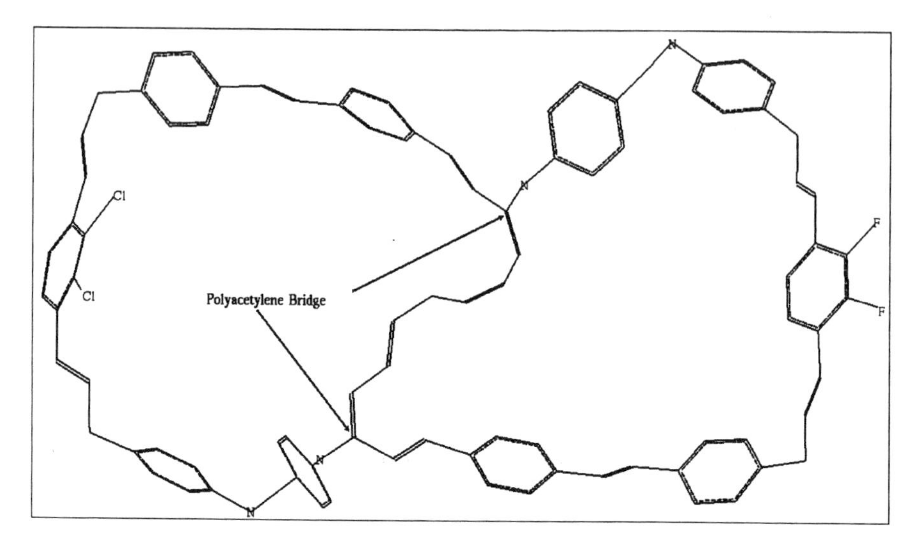

Figure 4. Wire diagram of nanologic circuit molecule without hydrogen atoms.

Once the energy-minimized structure was obtained, it was used as a starting point for molecular dynamics (MD) calculations. We were interested in performing MD simulations on this molecule in order to assess the amount of flexibility that it possesses. In order for the molecule to perform as already described the polymers from which it is constructed have to stay fairly rigid and the molecule as a whole should be planar. MD simulations allowed us to observe the behavior of the system in vacuum at room temperature. The molecule was simulated in vacuum, using a timestep of 1 femptosecond - which corresponds to the period of a hydrogen bond stretch. The system was first heated from absolute zero to 298 K over a heating run of 50 picoseconds, where the temperature of the system was increased by 3 K every 500 timesteps. Then the system was equilibrated for another 10 picoseconds to allow the energy to dissipate evenly throughout the system. At this point the system was ready for a constant room temperature dynamics simulation. The dynamics simulation was run for 50 picoseconds, again with a timestep of 10^{-15} seconds. The system maintained its shape fairly well through approximately 25 picoseconds of dynamics. But after 25 picoseconds of dynamics the molecule started to fold over on itself (Figure 5). This folding over process continued during the next 25 picoseconds of dynamics to yield the conformation depicted in Figure 6.

One of the reasons for running the dynamics calculations was to observe how flexible the molecule is. The electronic properties that were described would only be observed if the molecule could maintain a fairly planar shape. Only a planar shape maintains enough π orbital overlap intramolecularly and in-between the polymers which allows conjugation and conductivity. The results as depicted in Figures 5 and 6 show that a single molecule is very flexible and that ways to impose planarity on the molecule need to be examined.

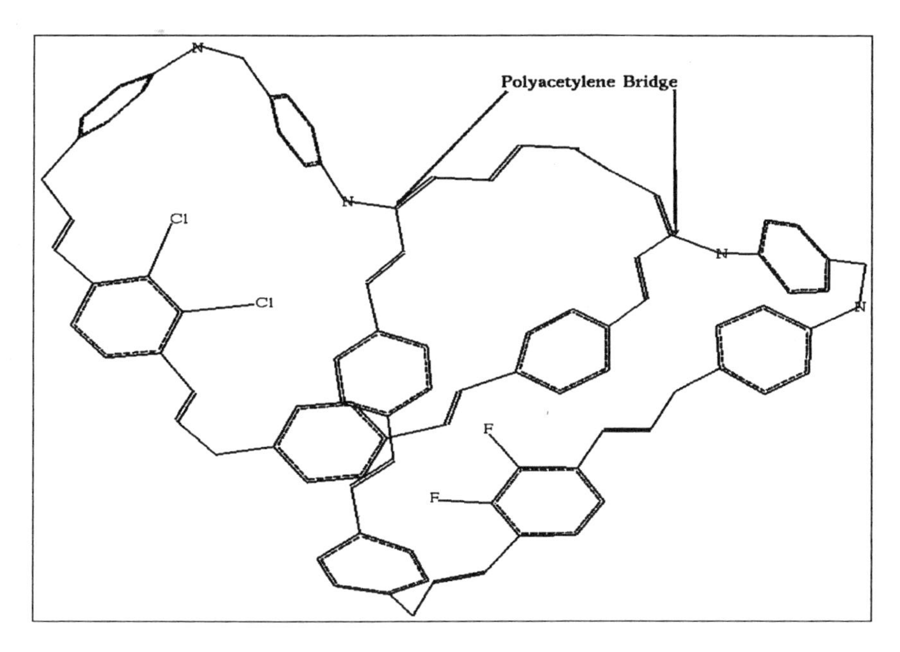

Figure 5. The nanologic molecule at 25 picoseconds.

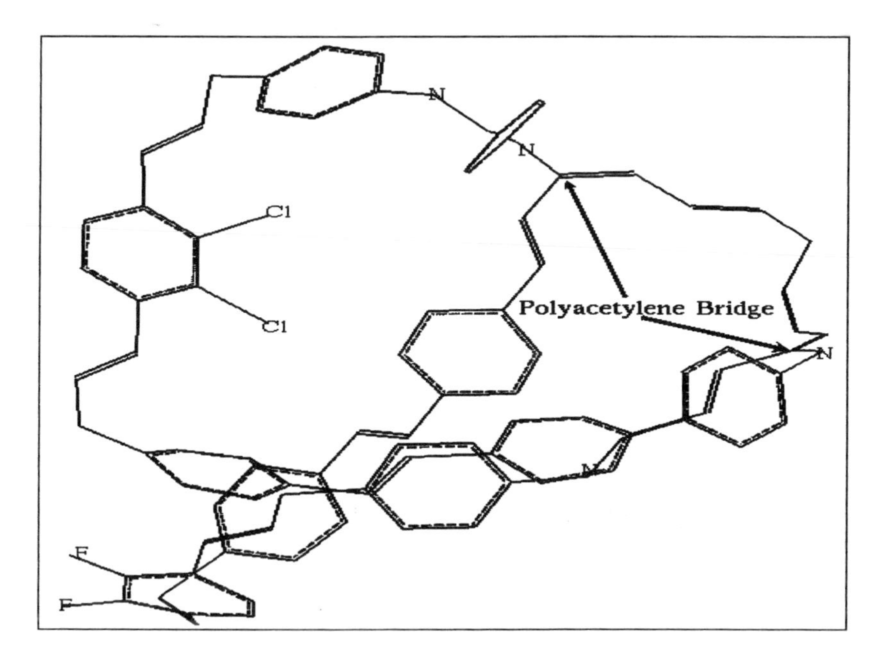

Figure 6. The nanologic molecule at 50 picoseconds.

Further information about the properties of the molecule was also obtained from the dynamics calculations. The frequency of oscillation at room temperature for each of the dihalo-substituted benzene rings was obtained by plotting each ring's rotational dihedral angle versus time. Figure 7 is a plot of the difluorobenzene ring to vinylene dihedral angle versus timestep for the duration of the 50-picosecond constant room temperature dynamics run.

Figure 7. Dihedral angle (in degrees) associated with the rotation of the difluorobenzene ring versus time in picoseconds (each tic represents 5 picoseconds).

The frequency of rotation of the difluorobenzene ring was determined to be 1.8 x 10^{12} Hz. We also obtained a similar plot for the rotational dihedral angle of the dichlorobenzene ring versus time. The results are found in Figure 8. For the dichlorobenzene ring, the frequency of rotation is estimated to be 3.6 X 10^{11} Hz. The difference in the calculated frequencies between the two dihalo-substituted benzene rings demonstrates that different frequencies of electromagnetic radiation are required to stimulate rotation for each.

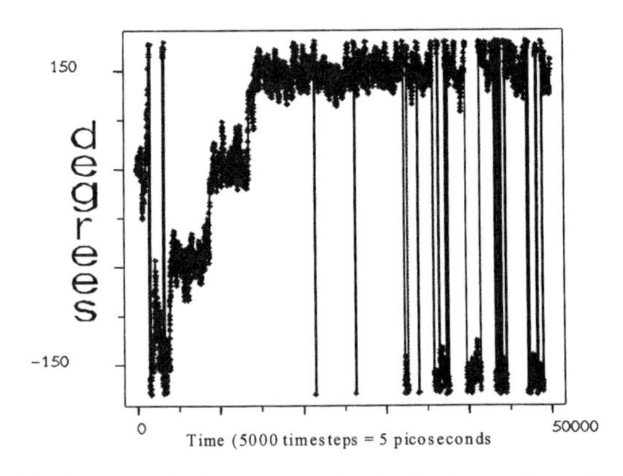

Figure 8. Dihedral angle (in degrees) associated with the rotation of the dichlorobenzene ring versus time in picoseconds (each tic represents 5 picoseconds).

As mentioned previously, due to the observed flexibility of the proposed nanologic molecule, means must be found to impose planarity on the molecule. One possible way of achieving this is by interconnecting more than one nanologic molecule together. This was achieved using Sybyl's molecular editing features, by attaching 6 carbon polyacetylene strands to the four corners of the molecule, and attaching another nanologic molecule to these strands. This molecule, which we will refer to as the dimer from here on, was also geometry optimized to give the lowest energy conformation using the AM1 semiempirical parameter set under Sybyl. The minimized dimer molecule is shown in

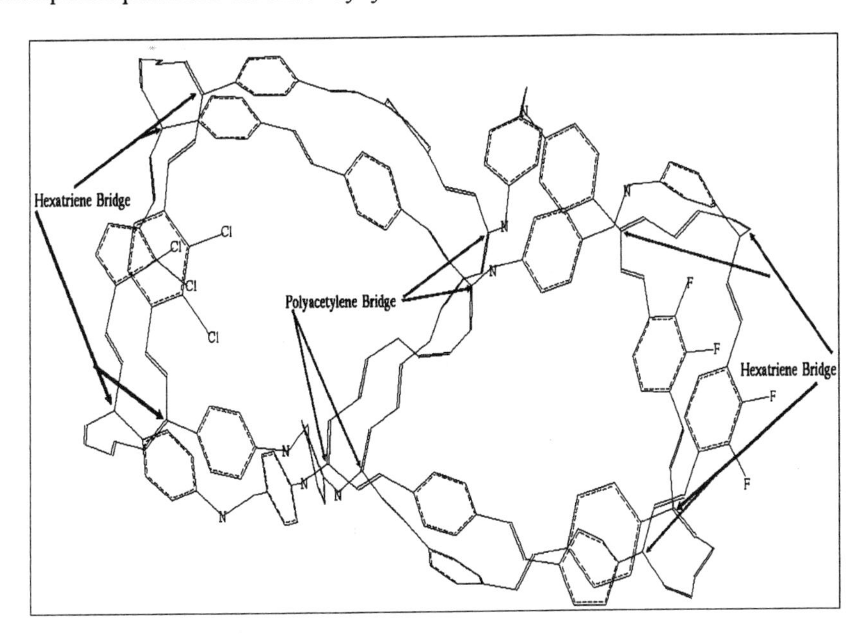

Figure 9. The minimized dimer of the nanologic molecule.

Figure 9. As we did previously for the 'monomer' of the nanologic molecule, this molecule was also used in a MD simulation. The dimer molecule underwent the same treatment as was described for the monomer molecule concerning heating, equilibration, and

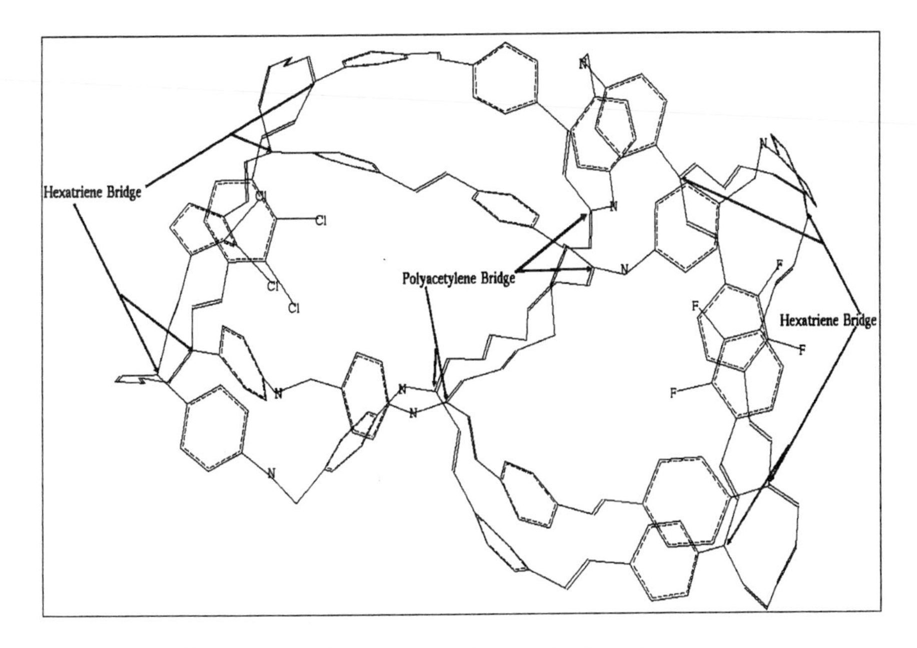

Figure 10. The dimer of the nanologic molecule at 25 picoseconds.

constant temperature runs. Again a 50 ps run was used for the production run. The results at 25 ps are shown in Figure 10, and at 50 ps in Figure 11. These two conformations are good representatives of most of the conformations of the dimer during

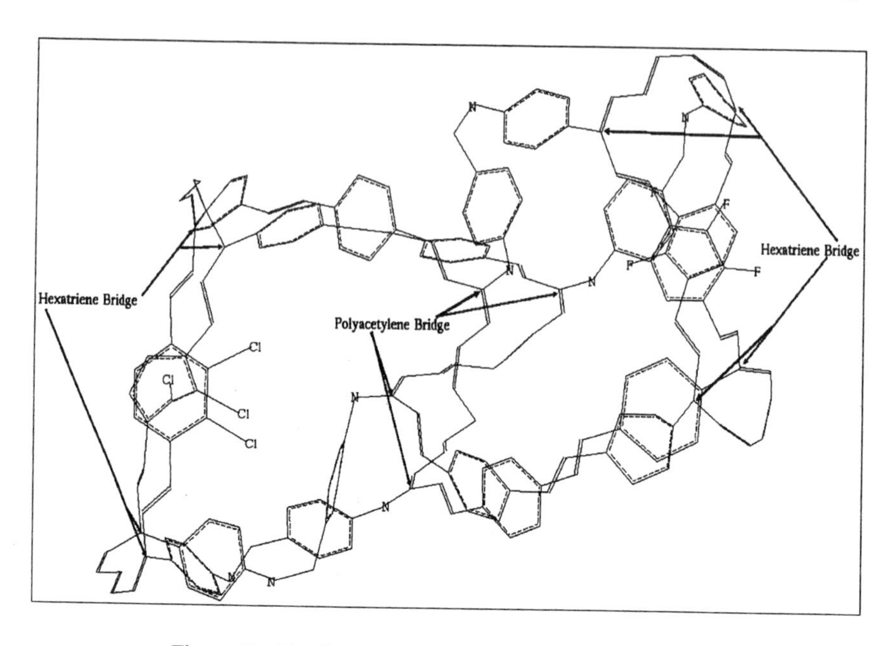

Figure 11. The dimer of the nanologic molecule at 50 picoseconds.

50ps of dynamics. The dimer showed that indeed the whole system had much less

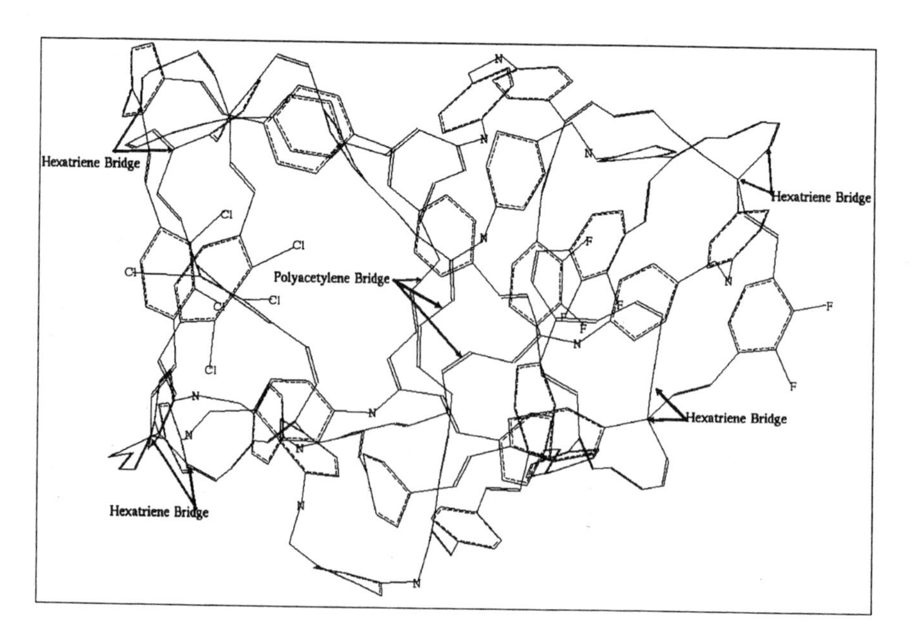

Figure 12. The trimer of the nanologic molecule at 25 ps.

flexibility than the monomer.

The success of the dimer molecule in reducing flexibility prompted us to explore the issue further by creating a 'trimer' molecule in the same fashion as the dimer. That is by using the same points of attachment, and the same 6 carbon strands of polyacetylene to attach a third nanologic molecule to the other two. The trimer molecule received the same treatment as the other two molecules regarding minimization, and dynamics under Sybyl. We looked at the conformations during the 50 ps production run, and once again the conformations at 25 ps and at 50ps are good representatives of the other conformations. The 25 and 50 ps conformations are found in figures 12 and 13 respectively.

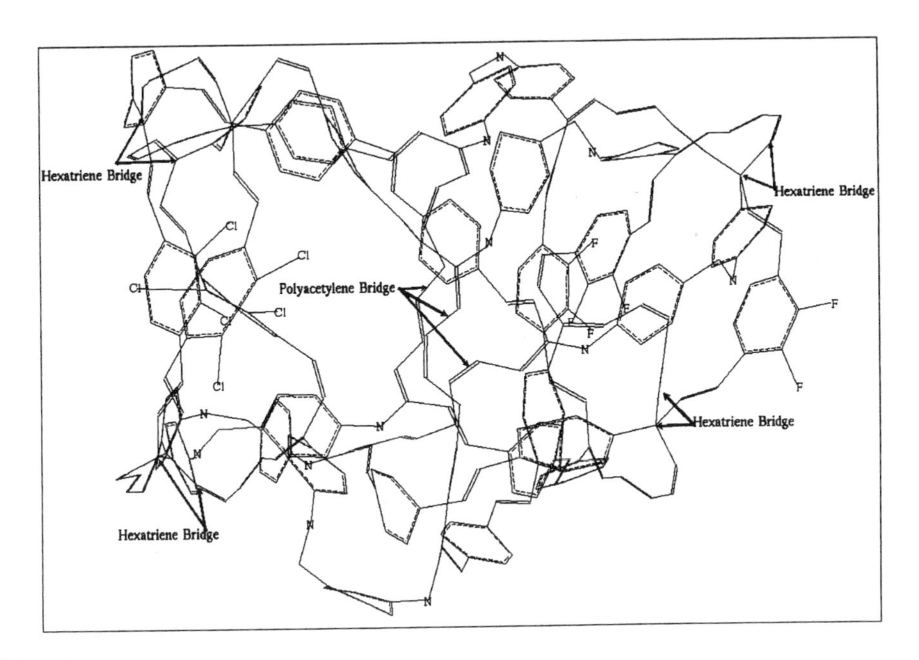

Figure 13. The trimer at 50 ps.

There is progressively less flexibility in the molecules when considering the progression from the monomer to the trimer. This suggests that the more nanologic molecules are assembled together, the less flexible the overall system, and as a result, the more planar each individual nanologic molecule is in the system. Planarity is essential for proper π orbital overlap, conjugation, and conductivity throughout each nanologic molecule.

CONCLUSIONS AND FURTHER CONSIDERATIONS

There is one additional modification that is needed to make practical circuits. The rotation of each benzene ring would induce an alternating current in the circuit due to the changing relationship between the loop (benzene ring) and the magnetic field. This can be thought of in terms of a commercial ac electrical generator. A solution to this oscillation in current is to use molecular rectifier molecules. Rectifier molecules like the ones proposed by Aviram and Ratner (A&R) could be used to convert the alternating current to direct current.[15] The placement of these rectifier molecules is shown in Figure 14. For details on the type of molecules that could be used and their proposed

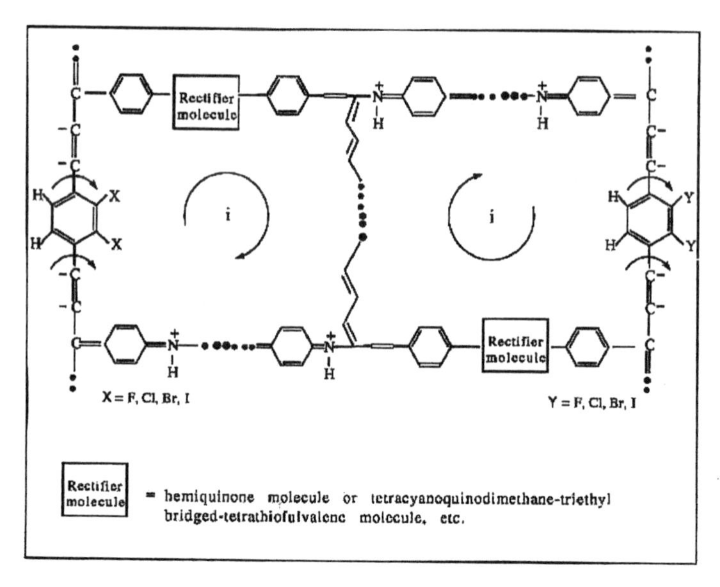

Figure 14. Modified nanologic circuit schematic showing placement of molecular rectifiers.

mechanism for rectification please refer to the A&R publication.

Planarity is crucial to the proposed electrical properties of the circuit. Linking more than one circuit together using a hexatriene molecule, appears to be an effective way to limit the motion of the molecule according to the results of our simulations, especially for three or more circuits. The fact that different frequencies of light can stimulate specific rings bearing specific substituents makes possible the creation of complex circuits (logic gates, counters, etc.) through the use of multiple interconnected nanologic circuits. According to our results, the more circuits that are connected together, the more rigid the structure will be, and the more planar each nanologic circuit molecule within the structure will be.

When considering other possibilities in maintaining planarity for the circuit, we suggest another alternative to linking multiple structures together. Another way to keep the molecule flat is to place it onto a substrate. The ideal substrate will be a non-conductive surface, which will allow the nanologic circuit to stick to it preferably through hydrogen bonding or other noncovalent interactions. Furthermore, the distance between the nanologic circuit and the substrate should not be so small as to interfere with the rotation of the benzene rings. After some time and consideration, we decided to use a sheet of boron nitride.[‡] A sheet of boron nitride is similar in structure to a sheet of graphite, with each boron being bound to 3 surrounding nitrogen atoms and each nitrogen being bound to 3 surrounding boron atoms. The lone pair electrons from each nitrogen atom can form hydrogen bonds with the hydrogen atoms on each of the polymers in the nanologic circuit molecule. A sheet of boron nitride (BN) was constructed and minimized using Hyperchem 5.01. The nanologic circuit molecule was then imported into Hyperchem in its geometry optimized state, and oriented and placed ~1.5 Å above

[‡] The use of boron nitride as a substrate was suggested to us by Dr.David Lindquist at the University of Arkansas at Little Rock.

the plane of the BN sheet. Then the whole structure was allowed to geometry optimize using the MM+ Molecular Mechanics parameter set, and the Polak-Ribiere optimization

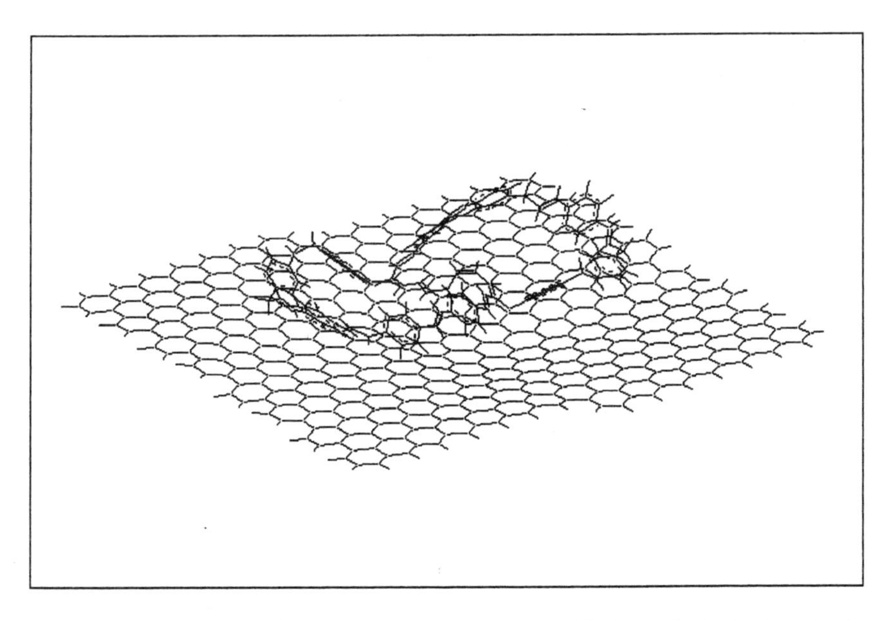

Figure 15. A sheet of boron nitride (542 atoms) as a substrate with the nanologic circuit molecule hydrogen bonded to it in the geometry-optimized conformation.

method under Hyperchem. A picture of this optimized system can be found in Figure 15.

Molecular dynamics simulations were performed using the optimized system as a starting point. The MM+ parameter set under Hyperchem was used. The system was started out at 0 K and allowed to slowly heat up to room temp (298 K), after which it was allowed to equilibrate. Production runs were performed only after the system was equilibrated. We were interested in three things: to observe the amount of bending that would occur in the system, second, to observe if there would still be enough room for rotation of the benzene rings, and third to see if the nanologic molecule would remain on the substrate. This research is still not yet concluded. However, preliminary results suggest that this substrate/circuit molecule system does not flex very much, and it seems as though the hydrogen bonding between the nanologic molecule and the nitrogen atoms on the BN sheet is strong enough to keep the molecule onto the substrate. It is not yet clear though whether the benzene rings still undergo proper rotation, and more work needs to be done to determine this.

ACKNOWLEDGEMENTS

The authors wish to thank: NASA Grant No. G-37020, Dr. Robert Shelton, NASA Johnson Space Center, Dr. Don W. Noid, and Dr. Bobby Sumpter of Oak Ridge National Laboratory.

REFERENCES

1. *The Wall Street Journal*, Tuesday, December 10, 1996.
2. Aviram, A., *J. Am. Chem. Soc.*, **1988**, 110, 5687.
3. Bumm, L.A., Arnold, J.J., Cygan, M.T., Dunbar, T.D., Burgin, T.P., Jones, L., Allara, D.L., Tour, J.M., Weiss, P.S., *Science*, **1996**, 271, 1705.
4. Samantha, M.P., Tian, W., Datta, S., Henderson, J.I., Kubiak, C.P., *Phys. Rev. B.*, **1996**, 53(12), 7626.
5. Wu, R., Schumm, J.S., Pearson, D.L., Tour, J.M., *J. Org. Chem.*, **1996**, 61(20), 6906.
6. Jones, L., II, Schumm, J.S., Tour, J.M., *J. Org. Chem*, **1997**, 62(5), 1388.
7. Tans, S.J., Devoret, M.H., Dai, H., Thess, A., Smalley, R.E., Geerlings, L.J., Dekker, C., *Nature*, **1997**, 386, 474.
8. Tour, J.M., Kosaki, M., Seminario, J.M., *J. Am. Chem. Soc.*, **1998**, 120, 8486.
9. The nanologic circuit concept was spawned out of discussions between the authors and D.W. Noid and B.G. Sumpter at Oak Ridge National Laboratory.
10. Drexler, K.E., *Nanosystems Molecular Machinery, Manufacturing, and Computation*, ©**1992**, John Wiley & Sons.
11. Nalwa, Hari Singh, Editor, *Handbook of Organic Conductive Molecules and Polymers*: 2, ©**1997** John Wiley and Sons.
12. *HyperChem 5.01*, ©**1997** Hypercube, Inc.
13. *Gaussian 94*, Revision E.1., M.J. Frisch, G.W. Trucks, H.B. Schlegel, P.M. W. Gill, B.G. Johnson, M.A. Robb, J.R. Cheesman, T. Keith, G.A. Peterson, J. A. Montgomery, K. Raghavachari, M.A. Al-Laham, V.G. Zakrzewski, J.V. Ortiz, J.B. Foresman, J. Cioslowski, B.B. Stefanov, A. Nanayakkara, M. Challacombe, C.Y. Peng, P. Y. Ayala, W. Chen, M.W. Wong, J.L. Andres, E.S. Replogle, R. Gomperts, R.L. Martin, D.J. Fox, J.S. Binkley, D.J. Defrees, J. Baker, J.P. Stewart, M. Head-Gordon, C. Gonzales, And J. A. Pople, Gaussian, Inc., Pittsburgh PA, **1995**.
14. *Sybyl* Version 6.3, ©**1995** Tripos Inc.
15. Aviram, A., and Ratner, M.A., *Chemical Physics Letters*, **29**(2), 277 (1974).

SHOCK AND PRESSURE WAVE PROPAGATION IN NANO-FLUIDIC SYSTEMS

Donald W. Noid, Robert E. Tuzun, Keith Runge, Bobby G. Sumpter
Chemical and Analytical Sciences Division
Oak Ridge National Laboratory
Oak Ridge, TN 37831-6197

INTRODUCTION

Sound is, by nature, a concept that is defined in continuous media. Typical treatments begin with the perturbation of a continuous fluid with some density and pressure [1,2]. These lead to a wave equation for pressure variations in which the speed of propagation (speed of sound) can be related to fluid density and other parameters. From these results and thermodynamic arguments other important quantities, such as the energy density of a sound wave, can be determined.

The smallest size scale so far at which sound has been experimentally observed is on the order of tenths of microns. For example, in a recent review Maris [3] discussed a "nanoxylophone", made of gold bars 150 atoms thick, which emitted sound waves with frequencies of about 8 THz for about 2 ns [4]. Within the last decade and a half, picosecond ultrasonic laser sonar (PULSE) [5] has begun to emerge as a technique for the investigation of thin multilayer films, most specifically thicknesses and interfacial properties [6]. Such studies are of crucial importance in the semiconductor industry.

Recent concerns about overly chaotic motion in classical systems [7] and the zero point energy problem [8] have led us to begin to explore the limitations of classical MD simulation. In the context of nano-fluidic systems, this would translate into whether or not it would be possible to study shock and pressure waves at the nanometer size scale using MD simulation. Of course, MD simulation has enabled reasonably accurate calculations of fluid viscosities [9]; however, the phenomenon of fluid viscosity, at the atomic level, does not require *coherent* energy transfer.

In this paper we report MD investigations of coherent fluid motion in a system we have previously simulated, namely helium inside a carbon nanotube. It is argued that MD simulation is unsuitable for studying such phenomena.

Simulation Methods

The MD simulations reported here are very similar to helium-carbon nanotube simulations in a previous paper in this journal [10]; only the differences will be described. Figure 1 shows the

Computational Studies, Nanotechnology, and Solution Thermodynamics of Polymer Systems
Edited by Dadmun *et al.*, Kluwer Academic/Plenum Publishers, New York, 2000

essentials of the system: a carbon nanotube which is static throughout the simulation and a region in which fluid atoms may move. The boundaries of this region are denoted z_1 and z_2, where z is the direction of the nanotube symmetry axis. The fluid atoms are initially placed in a hexagonal close packed lattice which does not approach within 3 Å of the nanotube. This entire system (fluid plus nanotube) is near mechanical equilibrium. In simulations in which the fluid atoms are initially motionless and no external forces are exerted on the system, the fluid temperature never rises above a few degrees K.

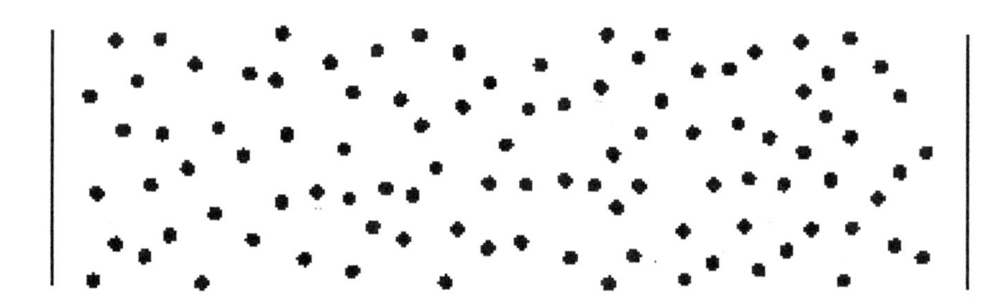

Figure 1 Sideways view of a fluid-carbon nanotube system. The solid lines represent the nanotube. The dashed lines represent either minimum image boundaries (shock wave simulations) or initial locations of driving and driven membranes (pressure wave simulations). Fluid atoms are enclosed within.

Two types of simulations, namely shock and pressure wave, are reported. In the shock wave simulations, z_1 and z_2 are fixed at the tenth and tenth to last rings of the nanotube and represent locations of minimum image boundaries described previously. A 10 Å plug of fluid is given an initial z velocity. Subsequent fluid motion is tracked very carefully through visualization and through extensive analysis.

In the pressure wave simulations, the boundaries represent idealized membranes located, respectively, at $z_1(t)$ and $z_2(t)$. These membranes have Lennard-Jones interactions with the fluid atoms with identical constants as the fluid-fluid interactions, but in the z direction only. The location of one membrane is varied sinusoidally

$$z_1(t) = z_{1o} + A\sin(2\pi t/\tau) \tag{1}$$

where z_{1o}, A, and τ are constants. The other membrane, in addition, has an interaction term of the form

$$V_2(z_2) = \frac{k}{2}(z_2 - z_{2o})^2 \tag{2}$$

and the force constant k and idealized membrane mass are chosen so that the vibrational period is τ.

Results and Discussion

The nanotube was kept static in each simulation reported here so that the motion of the nanotube would not perturb the fluid flow. In other words, conditions were made as favorable as possible for coherent energy transfer. Also, the system was equilibrated for 2 ps before the introduction of any shock waves or external forces. The flow region was about 100 Å long;

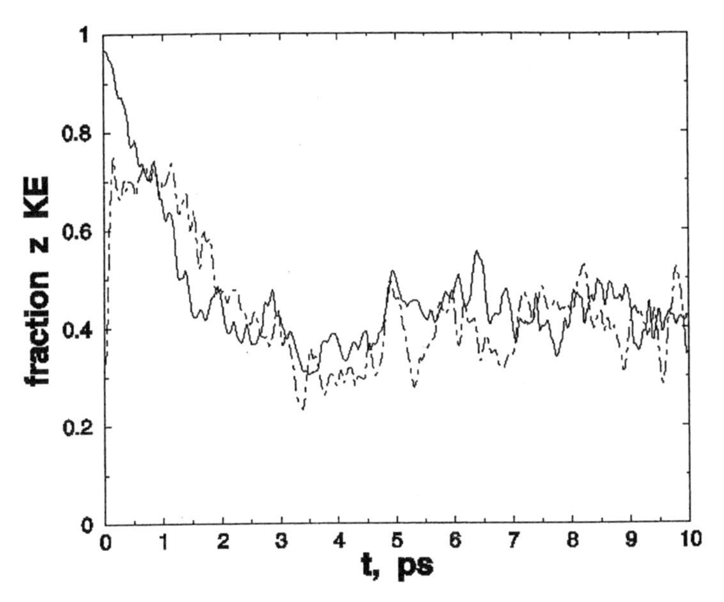

Figure 2 Typical shock wave simulation results (fraction of kinetic energy in the direction of flow) for helium within a 19.9 Å diameter nanotube. Solid line: central region atoms. Dashed line: adjacent region atoms.

nanotube diameters ranged from 16 Å (259 helium atoms inside the nanotube) to 32 Å (2157 helium atoms or ~.1g/ml).

Shock wave simulations were performed in which the 10 Å slice of fluid at the center of the nanotube was given an initial velocity of 10 Å/ps (near the speed of sound in helium at 100 K). The number of fluid atoms in this plug ranged from 25 for the smallest radius nanotubes to 211 for the largest. The initial fluid temperature (from random initial momenta) was about 100 K. Two groups of atoms were singled out for analysis: those in the central and adjacent fluid slices at the beginning of the simulation. The fraction of kinetic energy in the z direction was followed in each group of atoms. Typical results are shown in Fig. 2. The fraction of z kinetic energy steadily decreases in the central region atoms, averaging about the thermal equilibrium value, 1/3, within less than 5 ps. The adjacent region atoms suddenly gain z kinetic energy within about the first 0.1 ps and then a decreasing trend similar to that for the central region atoms. No coherent transfer of z kinetic energy is observed. In short, the motion very rapidly thermalizes, which is consistent with the results of visualization. In several simulations, positions of every fluid atom were saved every 0.05 ps. The movies showed that within 2 ps, the atoms in the initial plug essentially interspersed themselves within the adjacent fluid volume and then randomly dispersed. This behavior was observed at each nanotube radius.

The purpose of the pressure wave simulations (the previously described moving membrane simulations) was to determine if molecular dynamics would allow the coherent transfer of mechanical energy across a fluid. In these simulations, if energy transmission is to occur, the membrane vibrational periods must be commensurate with the length of the nanotube and speed of transmission.

Typical results are shown in Fig. 3, for which the bottom membrane vibrates with an amplitude of 0.5 Å and a period of 1 ps. An amplitude of 0.5 Å was chosen because at larger amplitudes, the fluid temperature built up, usually within 50 ps, to several thousand degrees. The top (driven) membrane, which has a mass of 40 amu and a fundamental vibrational period of 1 ps, shows no coherent motion even after 100 oscillations of the driving membrane. Other simulations, in which the period of vibration was up to 5 ps and the mass of the driven membrane up to 100 amu, yielded virtually identical looking results.

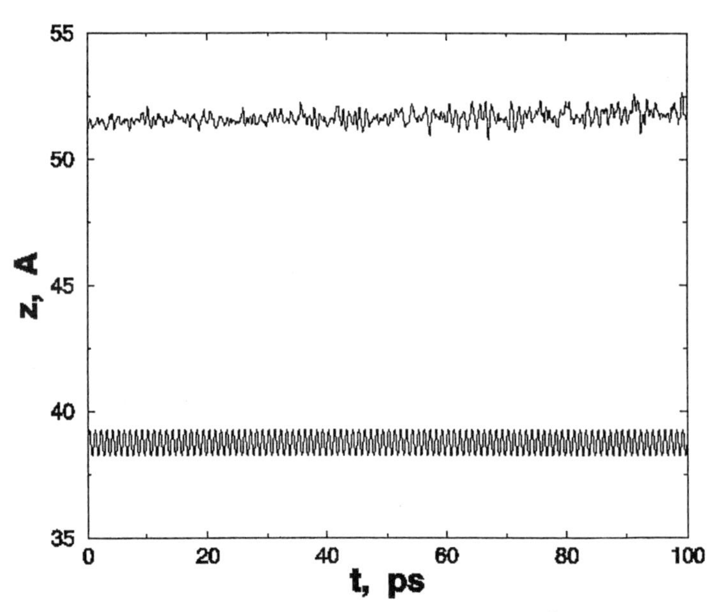

Figure 3 Typical pressure wave simulation results for helium within a 19.9 Å diameter nanotube: locations of driving (solid line) and driven (dashed line) membranes. The coordinates of the driving membrane are displaced 90 Å for the sake of easy visual comparison with the driven membrane coordinates.

The comparison of classical and quantum mechanical results, beginning 25 years ago with few atom systems, has so far yielded two important drawbacks: the zero point energy problem and chaos. The topic of classical-quantum correspondence has been recently studied in single polyethylene chains in vacuum. Calculations from the recently developed internal coordinate quantum Monte Carlo (ICQMC) method [8, 11] indicate that the ground vibrational state of polyethylene is extremely stiff; the probability distribution of end to end distance in a 100 monomer unit chain lies within a range of several tenths of an angstrom and is centered on the equilibrium planar zigzag value. In classical calculations at the quantum ground state energy, the polymer chain coiled; in other words, the flow of energy which quantum mechanically should have been locked in place was unrestricted in classical simulations. Even at energies at a fraction of the ground state energy, classical simulations showed large excursions in end to end distance and other indicators of overall positional stability. Classical simulations at very low energies (in which the temperatures reached only 2 K) showed high dimensional chaos [7].

The polyethylene system, a loosely connected bond network, was chosen as a worst possible case in order to illustrate the limitations of classical MD simulation. Systems with external constraints (such as nearby chains in a crystal) or cyclical bond networks would be expected to exhibit significantly restricted motion. In fact, classical simulations of carbon nanotubes, which have a two-dimensional bond network, showed a sizeable but significantly smaller disagreement with quantum results.

Of course, many of the essential features of a liquid are preserved in classical simulations. What determines these liquid properties more than the total energy is energy differences with respect to thermal motion, external forces, and so on. In the context of classical-quantum correspondence, it is important to note that any liquid is an unbound system and so the zero point energy problem has far less significance in considerations of liquid structure than in the problems previously discussed. However, for cases of coherent motion such as superfluidity, chaos is still an important consideration, and MD simulation will likely fail to describe these phenomena.

Conclusions

In two types of helium-nanotube setups, no coherent transmission of mechanical energy occurred in classical MD simulations. The fluid motion thermalized too rapidly to allow this to happen. This occurred despite the fact that conditions were made as favorable as possible for energy transmission (low temperatures, static nanotubes, near liquid densities).

We emphasize that MD simulation has worked well to predict some fluid properties such as viscosity and diffusion coefficient. However, it appears to be unsuitable for the study of phenomena requiring coherent fluid motion. Because of the inherent mechanical constraints, classical MD treatments may suffice for similar studies in solid nanostructures, depending on the degree of external mechanical constraints.

To what extent shock and pressure wave propagation occurs in nano-fluidic systems cannot be determined by classical simulation, nor has it been studied experimentally. If, in the future, such behavior were observed, a quantum mechanical treatment or some other method would be required for explanation.

Acknowldgments

Research sponsored by the Division of Materials Sciences, Office of Basic Energy Sciences, U.S. Department of Energy under contract DE-AC05-96OR22464 with Lockheed-Martin Energy Research Corp. Also supported in part by an appointment to the ORAU Postdoctoral Research Associates program administered jointly by the Oak Ridge Intitute for Science and Engineering and Oak Ridge National Laboratory.

References

[1] Goldstein H 1980 *Classical Mechanics, 2nd Ed.* (Reading: Addison-Wesley)

[2] Symon K R 1971 *Mechanics, 3rd Ed.* (Reading: Addison-Wesley)

[3] Maris H January 1998 *Scientific American* 86.

[4] Lin H-N, Maris H J, Freund L B, Lee K Y, Luhn H, and Kern D P 1993 *J. Appl. Phys.* **73** 37

[5] Morath C J, Collins G J, Wolf R G, and Stoner R J 1997 *Solid State Technology* **40** 85

[6] Tas G, Stoner R J, Maris H J, Rubloff G W, Oehrlein G S, and Halbout J M 1992 *Appl. Phys. Lett.* **61** 1787

[7] For a general discussion and an overview of existing methods, see Y. Guo, D. L. Thompson, and T. D. Sewell, J. Chem. Phys. 104, 576 (1996); T. D. Sewell, D. L. Thompson, J. D. Gezelter, and W. H. Miller, Chem. Phys. Lett. 193, 512 (1992).
J. M. Bowman, B. Gazdy, and Q. Sun, J. Chem. Phys. 91, 2859 (1989);
W. H. Miller, W. L. Hase, and C. L. Darling, 91, 2863 (1989);
D. A. McCormack and K. F. Lim, 106, 572 (1997);
M. Ben-Nun and R. D. Levine, 105, 8136 (1996).
R. Alimi, A. García-Vela, and R. B. Gerber, J. Chem. Phys. 96, 2034 (1992).
Newman D E, Watts C, Sumpter B G, and Noid D W 1997 *Macromol. Theory Simul.* **6** 577

[8] Kosloff R, Rice S. A., 1981 *J. Chem. Phys.* **74** 1340

Noid D W, Tuzun R E, and Sumpter B G 1997 *Nanotechnology* **8** 119
[9] Allen M P and Tildesley D J 1987 *Computer Simulation of Liquids* (Oxford: Clarendon Press)
[10] Tuzun R E, Noid D W, Sumpter B G, and Merkle R C 1996 *Nanotechnology* **7** 241 and references therein
[11] Tuzun R E, Noid D W, and Sumpter B G 1996 *J. Chem. Phys.* **105** 5494

INDEX

Aggregation number, 49, 50, 52

Berthelot technique (for negative pressure), 5
Biphasic interface, 69, 70, 74
Block copolymer micelles, 49
Boron nitride, 168

Calcium binding proteins, 127
Calcium binding site, 127
Calmodulin, 128, 129, 136
CED: *see* Cohesive energy density
Chain configuration, 99
Cohesive energy density, 29
Compatibilization of polymer blends, 69
Compatibilizers, 70, 76
Compressive modulus, 101
Concentration fluctuations, 37
Confinement effects, 93
Connected bond networks, 153, 156
Connection table, 153
Cooperative diffusion coefficient, 20
Copolymer sequence distribution, 71
Copolymers, 74
Core shell model, 48
Correlation functions, 40
Correlation length at theta-condition, 18
Correlation radii, length (DLS and Sans), 10, 16, 17, 18, 21, 23
Correlations, intra- and inter-molecular, 17
Coupling of mass and thermal diffusion, 42
Critical demixing, 9
Critical exponents, 23, 24
Critical micelle density, 52
Crossover, 23, 24

Debye form factor, 18
Demixing isopleth, 4
Density profile, 119
Dezymer binding site, 130, 137
Diffusion, 83, 87, 175
Diffusion theory, 43
Dispersive interactions, 36
Domain size resolution, 85

Double critical points: *see* Hypercritical pressure; Hypercritical temperature
Dynamic correlation length, 20
Dynamic light scattering (DLS), 9, 19

EF-hand proteins, 137
Electron density distribution (EDD), 140
Electronic properties, 108
Escherichia coli (E. coli), 130

Fibers, 120–123
Flory interaction parameter, 57
Flory-theta temperatures, 1
Fluctuation hydrodynamics, 38

Glass transition, 100, 188

Hartee–Fock, 140, 141
High concentration labeling method, 17
Hydrophobic, 130
Hypercritical pressure, 4
Hypercritical temperature, 4

Interfacial modifiers, 70, 74
Internal coordinate Quantum Monte Carlo, 151–154, 174
Internal pressure, 30
Ising model, 23
Isotope effects on phase equilibria, 8

Kinematic viscosity, 43
Kinetics of phase separation, 61

Laminar electronic flow, 140, 142
Light scattering in PS/toluene, 39
Liquid–liquid demixing, 1
Lower critical solutions temperature, 57

Mass diffusion, 41
Master curve (surface tension), 31
Melting point, 100, 188
Metastable region, 60
Metropolis sampling, 71, 118
Microdroplets, 80

Microwave radiation, 160
Mie Theory, 81, 82, 93, 108
Miscibility, 56
Molecular basis of diseases, 127
Molecular dynamics simulations, 94, 98, 108, 151, 171, 175
Molecular electronic logic circuit, 159
Molecular orbitals, 145
Monte Carlo simulation, 70
Multiple hypercritical points, 6

Nanofiber stability, 121
Nanofluids, 171
Nanologic circuit, 168
Nanologic molecule, 165, 167
Nanoparticles, 93, 98, 107
Nanotechnology, 139, 151, 171
Nanotube, 153, 155, 171, 172
Negative pressures, 3
Neutron contrast factor, 47
Neutron scattering length density, 47
Non-equilibrium enhancement, 40

Optical beam bending technique, 41
2-D optical diffraction, 79
Ornstein–Zernicke structure factor, 18

Parvalbumin, 136
Pauli principal, 140
Phase diagram, 16
Phase equilibria, 2
Phase separation in polymer blends, 72, 74
Poincare surface of section, 108
Poly(methyl methacrylate), 70, 75
Poly(vinyl pyridine), 70, 75
Polyacetylene, 160, 167
Polyethylene, 94, 96, 118, 155, 174
Polyethylene oligomers, 33
Polymer blend, 79
Polymer composite, 79
Polymer films, 117, 119
Polymer overlap concentration, 19
Polymer particles, 80, 93, 96, 108
Polypropylene, 118
Polystyrene, 70, 75
Polystyrene/solvent phase equilibria, 1–11
Pressure-induced phase separation, 56, 66
Pressure wave, 171, 173
Proteins, 127
Pseudo-energy, 130

Quantum drops, 107, 108, 111

Radius of gyration, 15, 17, 21, 23, 48, 71, 107, 108, 111, 121
Rational design of novel proteins, 127
Reactivity ratios, 74
Rectifier molecules, 167, 168
Reduced phase diagrams, 7
Relaxation rate, 20
Reptation technique, 71
Resonance tunneling diode, 143, 146, 147
Rotational isomeric state model, 118, 123

Scaling representations of demixing isopleths, 11
Second virial coefficient, 22, 46, 48
Semiconducting wire, 146
Shock wave, 171, 173
Simulation methods, 152
Small angle neutron scattering (SANS), 9, 19, 46
Solubility parameter, 29
Solvent quality, 16
Soret effect, 38
Spinodal decomposition, 64
Stark effect, 111
Supercritical fluid, 15, 21, 23, 25, 45
Supra-molecular chemistry, 139
Surface effect, 98, 110
Surface energy, 119, 120
Surface tension, 31
Surfactants, 45

Temperature composition diagram, 58
Thermal diffusion, 37
Thermal diffusion coefficient, 42
Theta solvent/poor solvent transition, 5
Theta temperature, pressure, 16, 18, 21, 23, 25
Time dependent correlation functions, 38

Universality class, 15, 23
Upper critical solutions temperature, 57

Viscosity, 20, 171, 175

Zero point energy problem, 151, 152, 155, 171, 176
Zimm plot, 47